Separation Needs for Energy Independence and Environmental Sustainability

Topical Conference at the 2010 AIChE Annual Meeting

Salt Lake City, Utah, USA
7-12 November 2010

ISBN: 978-1-61782-136-3

Printed from e-media with permission by:

Curran Associates, Inc.
57 Morehouse Lane
Red Hook, NY 12571

Some format issues inherent in the e-media version may also appear in this print version.

Printed by Curran Associates, Inc. (2011)

For permission requests, please contact AIChE
at the address below.

AIChE
3 Park Avenue
New York, NY 10016-5991

Phone: (203) 702-7660
Fax: (203) 775-5177

www.aiche.org

Additional copies of this publication are available from:

Curran Associates, Inc.
57 Morehouse Lane
Red Hook, NY 12571 USA
Phone: 845-758-0400
Fax: 845-758-2634
Email: curran@proceedings.com
Web: www.proceedings.com

TABLE OF CONTENTS

(176a) Membrane Technology– POTENTIAL for A LOWER CARBON Future

Anthony Fane, UNESCO Centre for Membrane Science & Technology, The University of New South Wales, Sydney, Australia

Following a brief retrospective of membrane technology, this presentation considers a major challenge and opportunity for the industry. It is now evident that membrane technology has a major future in water and wastewater processing, including low pressure membranes (MF and UF) for water treatment, membrane bioreactors (MBRs) and as pretreatment, and high pressure RO is the preferred technology for desalination and water reclamation. However membrane technology has an image of being 'energy expensive' and this poses a major challenge with the need to curb green house gases. This presentation examines the role of membranes in the water industry and the possible responses to climate change. It is based on assessments of global technology trends as well as examples of research findings from the speaker's laboratories in Singapore and Sydney.

Seawater desalination by reverse osmosis (RO) is growing by at least 15% per year, and a 'business as usual' approach could see SWRO becoming a major GHG emitter by 2050. Reduced energy will come from developments in membranes, modules and operation. The prospects for very high permeability RO membranes requires redesign of the RO cascade with energy savings of 30 to 40% possible but at a capital cost penalty. The spiral wound RO element is amenable to further improvement as shown by CFD analysis. Significant energy savings, with GHG reduction, can be achieved by improved control of fouling in RO desalination and reclamation processes.

Membrane alternatives to RO include membrane distillation (MD) and forward osmosis (FO). Both offer lower GHG strategies. MD is a thermal process but uses available waste heat or solar energy and little electrical energy. It can operate effectively at very high recovery (product/feed) and can be coupled with a crystallizer to produce salt solids, and may be viable as a 'concentrate' treatment process. Water production from waste water by a dual membrane process (MF/UF + RO) comes at about half the energy cost of RO desalination. However the pretreatment energy now becomes significant and increases with flux. LCA shows that to minimize energy requires a much lower flux than the economic minimum; again a capital cost penalty to reduce GHG.

The MBR shows dramatic growth in application. Energy savings are being achieved by optimization of air-scour, and research into complex membrane fouling. Anaerobic MBRs are being developed and could also achieve energy savings. Membrane contactors could upgrade biogas. Novel MBRs incorporating MD and/or FO and driven by waste or solar heat could become alternatives to the conventional MBR + RO reclamation process , with a smaller GHG footprint.

Finally, membranes provide an opportunity for generation of renewable power based on the controlled mixing of salt water and fresh water. One approach uses pressure retarded osmosis, which has a long history but has been revived as a possible answer to reducing GHG emissions. This process promises a source of renewable energy, but also presents formidable challenges to the membrane community in terms of membrane and module design, pretreatment and fouling control.

See more of this Session: Session in Honor of Professor Anthony Fane: Membrane Separations for Sustainable Water and the Environment
See more of this Group/Topical: Topical 1: Separation Needs for Energy Independence and Environmental Sustainability

(176b) Membrane Aeration a Review

Michael J. Semmens, Civil Engineering, University of Minnesota, Minneapolis, MN

Currently wastewater treatment is extremely energy intensive and by some estimates accounts for approximately 2% of the electricity use in the USA. The installation of membranes in wastewater treatment plants to improve treated water quality tend to increase energy consumption. Novel approaches to reduce energy consumption in wastewater treatment are urgently needed. This paper will explore the benefits of membrane aeration for supplying biological processes with oxygen . The paper will also include discussion of the salient design issues and benefits as well as the challenges for implementation.

See more of this Session: Session in Honor of Professor Anthony Fane: Membrane Separations for Sustainable Water and the Environment
See more of this Group/Topical: Topical 1: Separation Needs for Energy Independence and Environmental Sustainability

DEVELOPMENT OF THIN-FILM COMPOSITE FO HOLLOW FIBER MEMBRANES AND THEIR POTENTIAL APPLICATIONS

R. Wang[1,2], *L. Shi*[2], *S.R. Chou*[1,2], *W.X. Fan*[1], *C.Y. Tang*[1,2], *C.Q. Qiu*[1,2], *A.G. Fane*[1,2]

1. Singapore Membrane Technology Centre, Nanyang Technological University, Singapore
2. School of Civil and Environmental Engineering, Nanyang Technological University, Singapore

Introduction

Forward osmosis (FO) process has gained much attention recently because of its potential to either reduce energy consumption (in seawater desalination) or produce energy (pressure retarded osmosis) [1]. However, the viability of FO technology for large-scale implementation has been impeded by two main technical challenges: (1) the identification of a suitable draw solution which is able to generate a high osmotic pressure and is readily regenerated; and (2) the lack of appropriate membranes for FO application that can produce a reasonably high flux. This paper describes the development of novel hollow fiber FO membranes with a focus on the effect of porous substrate on the performance of resultant FO membranes. Five types of UF hollow fiber substrates with distinctive structures were purposed designed and fabricated by carefully adjusting the polymer dope formula and spinning conditions. Subsequently, an ultra-thin polyamide-based RO-like selective layer was made on the UF-like inner surface of the substrates using the same condition of interfacial polymerization. With the aid of membrane characterization by a series of standard protocols, the effects of substrate structure on the formation of the selective layer and FO performance of the final membranes are explored.

Experimental

Commercial polymer polyethersulfone (PES) was used for various hollow fiber substrate preparation using the dry-jet wet spinning method. The formation of a RO-like skin layer on the inner surface of the PES hollow fiber substrate was made based on interfacial polymerization using m-phenylenediamine (MPD) aqueous solution and trimesoyl chloride (TMC) hexane solution.

The hollow fiber substrates and FO membranes were characterized by a series of standard protocols in terms of membrane morphology, structure (fiber dimension, mean pore size, molecular weight cut off (MWCO), and porosity), mechanical strength, water permeability and salt (NaCl) permeability. In FO experiments, sodium chloride (NaCl) was used as the draw solute, and the feed water was de-ionized (DI) water or DI water containing 50 ppm Dextran with molecular weights of 200~300 kDa. FO experiments were performed in two configurations: active layer facing draw solution (AL-facing-DS) and active layer facing feed water (AL-facing-FW). The volumetric flow rates of the shell side and lumen side were 1500 mL/min and 450 mL/min, respectively, to ensure a similar Reynolds number (around 2600) of the liquid flowing both in the module shell and fiber lumen. The volumetric water flux, J_V, was determined at a certain time interval by measuring the weight changes of the feed tank with a digital mass balance connected to a data logging system. The FO test was conducted at room temperature of 23 °C. The experimental details can be found in literature [2].

Results and discussion

Structure and morphology of PES hollow membrane substrates

Figs. 1-5 show the cross-section and surface morphologies of the morphology of PES hollow fiber substrates with different structures, and Table 1 shows the characteristic features of the membranes. It can be seen from Fig. 1 and Fig. 2 that #A and #B hollow fiber substrates exhibited similar cross-section morphologies. The major difference between #A and #B substrate lies in the MWCO of the outer layer. As shown in Table 1, the MWCO data tested from shell side for #A was smaller (24 kDa) than that of #B (88kDa), which suggested that #A substrate possessed a denser UF outer skin on the shell side.

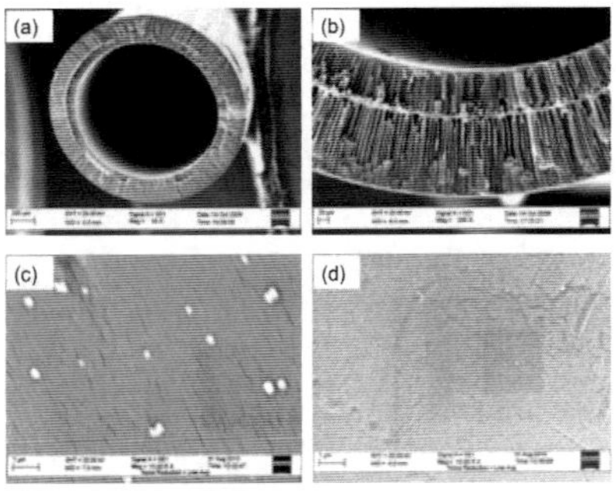

Fig. 1. Morphology of #A PES hollow fiber substrate. (a) cross-section at 45x, (b) partially enlaged at 200X, (c) inner layer surface at 10K X, and (d) Outer layer surface at 10K

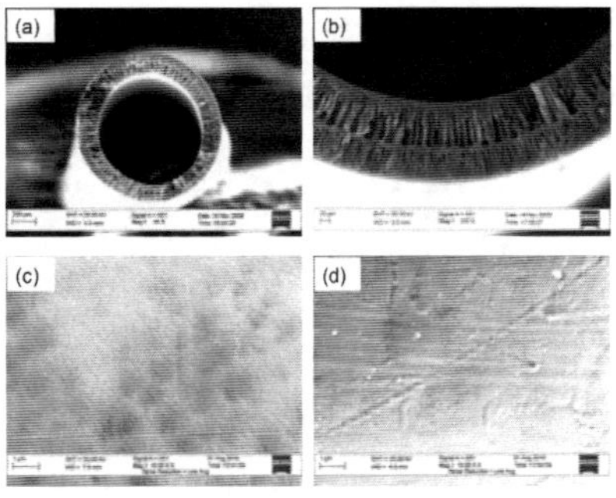

Fig. 2. Morphology of #B PES hollow fiber substrate. (a) cross-section at 45x, (b) partially enlaged at 200X, (c) inner layer surface at 10K X, and (d) Outer layer surface at 10K

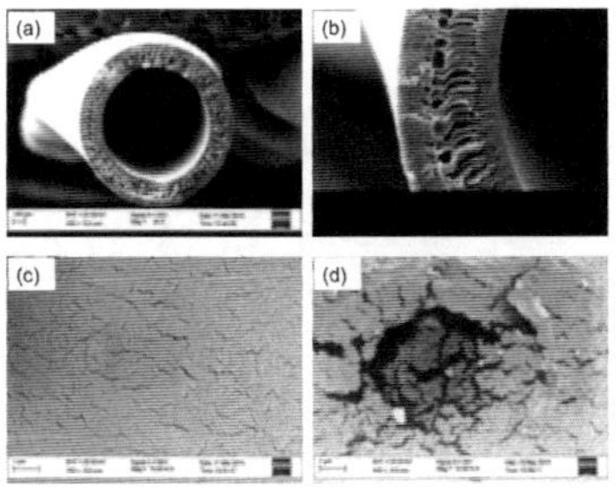

Fig. 3. Morphology of #C PES hollow fiber substrate. (a) cross-section at 45x, (b) partially enlaged at 200X, (c) inner layer surface at 10K X, and (d) Outer layer surface at 10K

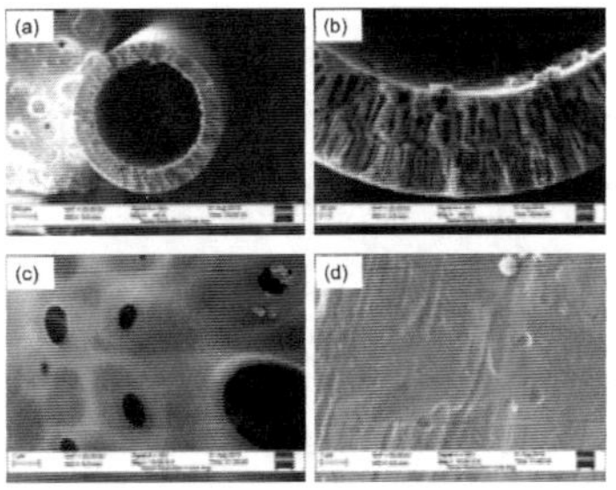

Fig. 4. Morphology of #D PES hollow fiber substrate. (a) cross-section at 45x, (b) partially enlaged at 200X, (c) inner layer surface at 10K X, and (d) Outer layer surface at 10K

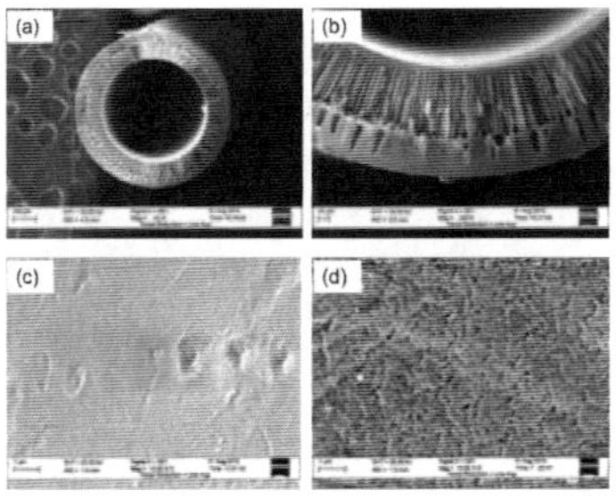

Fig. 5. Morphology of #E PES hollow fiber substrate. (a) cross-section at 45x, (b) partially enlaged at 200X, (c) inner layer surface at 10K X, and (d) Outer layer surface at 10K

It was found from Fig. 3 that needle-like pores were developed from both the inner and outer surface, and a layer of macro-voids with thin cell walls was sandwiched in between. The porosity of the substrate fiber was as high as 82% due to the existence of the macro-void. When measured from the shell side of the hollow fiber substrates, the substrate showed a higher MWCO of 520 kDa and a high pure water flux of 278 L/m^2.h.atm similar with that of #A and # B. It suggested that the outer UF-like skin has been successfully avoided as originally intended.

The #D substrate was deliberately designed to distinguish from the other substrates. As shown in Fig. 4, the compact needle-like pores underneath the outer layer were developed tightly towards inside until the middle depth of cross-section, macrovoids and sponge-like structure were formed near the inner side. The porosity of #D substrate was as low as 63%, much smaller comparing to #A, B #C and #E, causing the lowest pure water permeation among all the substrates.

The morphology of the #E substrate looks like an inverse image to the #D substrate, where long needle-like pores were developed from the inner side underneath a thin sponge like layer that was close to the inner surface, and a big sponge-like portion was located at the outer side of the substrate wall. Besides, the MWCO rejection from the shell side was 380 kDa, but no huge cavities existed there as shown in Fig. 5d.

Based on the MWCO results obtained from the lumen side, all the mean pore size and standard deviation of hollow fiber substrate were calculated and also included in Table 1. The method for the calculation can be found in literatures [3]. It was found that the #A, #B and #C three substrates possessed a narrow pore size distribution as suggested by small geometric standard deviations of 1.07, 1.08, and 1.15 nm, respectively. Among all the substrates, the #D had the largest mean pore size 24.37 nm and the most wide pore size distribution (standard deviation as high as 1.25). This result is not surprising as referring to the SEM pictures shown in Fig. 4. The suitability of the #D substrate for subsequent thin film formation on its top will be discussed in next section.

Table 1. Characteristics of PES substrates

Sample	Dimension		Mean pore size (lumen) D (nm)	Standard deviation[#] s (nm)	PWP (L/m²h概ar)	MWCO (kDa) (shell)	Porosity (%)
	ID(祄)	Thickness (祄)					
#A	1060	190	19.5	1.07	280	24	84
#B	960	180	12.7	1.08	275	88	75
#C	960	205	9.6	1.15	278	520	82
#D	960	180	24.4	1.25	50	67	63
#E	900	180	8.3	1.23	130	380	69

Effect of substrate structure on the formation of the selective layer

The RO-like selective layer attached to the inner skin of the hollow fiber substrate was fabricated using interfacial polymerization method, and the all the conditions were kept the same. The performance data of the hollow fiber membranes in RO mode are listed in Table 2. The membranes #A, #B, #C and #E show high salt rejection (R_s>90%), and high water permeability, which indicates that the TFC skin of these membranes can serve well as a semipermeable barrier. However, the #D membrane shows relatively high water permeability but poor salt rejection of 15.2% to NaCl, which is unsuitable to be used as an osmosis barrier. Since the operating conditions of interfacial polymerization used were the same, obviously, the property of the inner surface of the substrate should be responsible for the quality of the thin film.

Table 2. Intrinsic separation properties of FO membranes

Sample	RO test		
	A (L/m^2.h.bar)	B (L/m^2.h)	Rs (%)
#A-FO	3.29	0.11	94.5
#B-FO	3.07	0.12	93.6
#C-FO	3.35	0.08	95.4
#D-FO	5.09	25.03	15.2
#E-FO	3.93	0.27	90.4

When performing the interfacial polymerization experiment, the substrate was firstly soaked in a MPD aqueous solution, which favorably tended to penetrate into the pores on the substrate surface if there existed. However, if the pore size was small, the surface tension would prevent the ammine solution from the penetration. Subsequently, the excess MPD aqueous solution was removed from the lumen of the hollow fiber substrate, and a thin MPD liquid film was left on the inner surface to continue the reaction with TMC in the following step.

As shown in Figures 1c~5c, the inner surface morphology of the #D substrate is different from the other substrates, where huge cavities with diameter around 2μm together with large holes with an approximate dimension of 1μm are visibly observed under a magnification of 10000 times instead of the tiny cracks on the surface of the other substrates. Thus, it was very likely that the thin MPD liquid film left on the #D substrate cannot maintain integrity with the removal of excess MPD solution, as the MPD aqueous solution can penetrate into these big holes. Consequently, after reacting with acid chloride, the polyamide layer formed on the lumen skin of the #D substrate might have defects in the region of big holes, leading to the poor salt rejection. A similar discussion was conducted by Singh [4].

Effect of substrate structure on FO performance

Table 3 gives the performance data (water flux and salt flux) of four FO hollow fibers using 0.5M sodium chloride (NaCl) as the draw solute, and de-ionized water as the feed at 23 ºC. The mass transfer coefficient ($K_m = D/S$) and membrane structural parameter ($S = \tau l/\varepsilon$, where l is substrate thickness, τ tortuosity and ε porosity), which were calculated based on FO solution-diffusion model and FO experiments [2], are also included in the table. Overall, the AL-facing-DS configuration shows much higher water flux than the AL-facing-FW configuration for all the membrane. This trend is attributed mainly to the concentrative/dilutive ICPs, which has been well studied in literature. The water flux of the four FO membranes was in the range of 43-49 L/m^2.h using 0.5 M NaCl as the draw solution and DI water as the feed for the AL-facing-DS orientation at 23 °C. The corresponding ratio of salt flux to water flux was <0.081 g/L, which is superior to all other FO membranes reported in the open literature under the same operating conditions.

Table 3. FO membrane performances using DI water as feed

Sample	FO test - AL facing Draw				FO test - AL facing Feed			
	Jv (L/m2.hr)	Js (g/m2.hr)	Km (L/m2.hr)	S (mm)	Jv (L/m2.hr)	Js (g/m2.hr)	Km (L/m2.hr)	S (mm)
#A-FO	47.7	3.5	9.25	0.63	18.6	2.0	11.8	0.49
#B-FO	49.4	3.9	10.02	0.58	16.7	1.2	11.1	0.52
#C-FO	43.6	2.8	9.13	0.63	18.7	1.6	11.9	0.49
#E-FO	49.0	4.0	9.93	0.58	17.6	1.2	11.3	0.51

It can be seen again from Table 3 that the four FO hollow fiber membranes show very similar performances in either the AL-facing-DS configuration or the AL-facing-FW configuration. This result is anticipated as the four membranes possess similar intrinsic separation properties, showed in Table 2. According to our previous study [5], if DI water and 0.5 M NaCl were used as the feed and draw solution, the resistance of the substrate is much less than the RO-like skin, and the water permeation through the RO-like skin is the rate-determining step.

To further understand the impact of the substrate structure on the FO performance of the final membrane in non-ideal situations, experiments were performed using DI water containing 50 ppm dextran with molecular weights of 200 ~300 kDa as the feed. Fig. 2 shows the water flux of the four FO membranes as function of operating time in the AL-facing-DS configuration. It is observed that for the four membranes, the initial high water flux around 43-49 L/m^2.h using DI water as the feed cannot be maintained with the addition of 50 ppm dextran in the feed stream. The water flux declined significantly in the initial two hours, and the decrease rate became gradually slow and finally reached a steady state after 22 hours. In addition, the difference in water flux profile for the four FO membranes can be identified clearly. The performance of the four membranes is in the sequence of #B>#A>#C>#E, i.e., the #B shows the highest water flux while the #E shows the lowest water permeation.

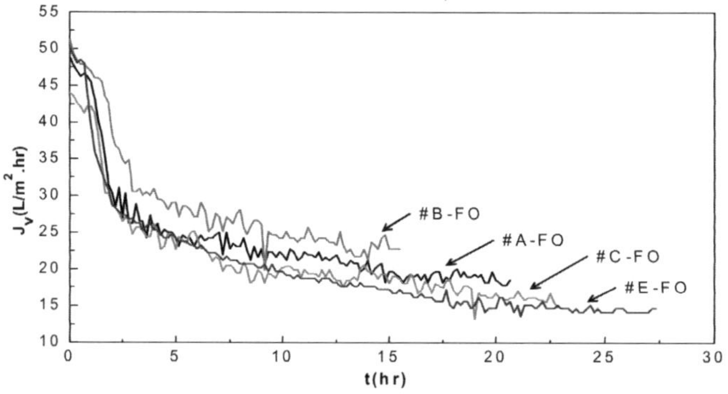

Fig. 2. Water flux in FO process in AL-facing-DS configuration (feed: DI water with 50 ppm daxtran; draw solution: 0.5 M NaCl)

Obviously, the FO membrane performance was affected by the substrate structure as the feed was in contact with the substrate. Referring to Table 1, #A and #B FO membranes have much tightened outer skin layers than #C and #E (24, 88, 520, 380kDa, respectively). As the foulants used are dextran with molecular weights of 200 ~300 kDa, obviously, they can penetrate into the porous substrates of #C and #E easily as compared to the substrates of #A and #B membranes. The existence of the foulant in the substrate can block the membrane pores and thus actually modified the substrate structure, which was reflected by the significant change of the apparent membrane structural parameter. The structural parameter of #C and E varied from ~0.6 to 1.4/1.5 mm, while the change of the structural parameter of #A and B was much smaller (~0.6 to 1.0/1.1 mm). It indicated that the surface structure of the FO membrane facing the feed is very important as it is directly associated with the tendency of FO membrane fouling.

Conclusions

The substrate surface structure is very important for fabricating a RO-like thin film. A substrate with mean pore size <20 nm with a narrow pore size distribution should be preferred to obtain a good semipermeable skin based on current study.

It may not be appropriate to use DI water and a low concentration draw solution (i.e, an ideal situation) to evaluate the FO membrane, as the water permeation through the RO-like skin is the rate-determining step, and the effect of the substrate structure cannot be fully reflected.

FO membrane can be fouled severely under certain circumstances. The surface structure of the FO membrane facing the feed is very important as it is directly associated with the tendency of FO membrane fouling.

Acknowledgements

We would like to thank the Environment and Water Industrial Development Council of Singapore for funding support under the project # EWI RFP 08/01. We are also grateful to Singapore Economic Development Board for funding Singapore Membrane Technology Center.

References

1. T. Y. Cath, A. E. Childress, M. Elimelech (2006), "Forward osmosis: principles, applications, and recent developments", *Journal of Membrane Science*, 281, pp.70.
2. R. Wang, L. Shi, C. Y. Tang, S. Chou, C. Qiu and A. G. Fane (2010), "Characterization of novel forward osmosis hollow fiber membranes", *Journal of Membrane Science*, 355, pp.158-167.
3. X. Wei, R. Wang, Z. S. Li and A. G. Fane (2006), "Development of a novel electrophoresis-UV grafting technique to modify PES UF membranes used for NOM removal", *Journal of Membrane Science,* 273, pp. 47.
4. S. P. S. Singh, S. V. Joshi, J. J. Trivedi, C. V. Devmurari, A. P. Rao and P. K. Ghosh (2006), "Probing the structural variations of thin film composite RO membranes obtained by coating polyamide over polysulfone membranes of different pore dimensions", *Journal of Membrane Science,* 278, pp.19-25
5. Chou, L. Shi, R. Wang, C. Y. Tang, C. Qiu and A. G. Fane (2010), " Characteristics and potential applications of a novel forward osmosis hollow fiber membrane", *Desalination,* 261, pp. 365-372

(176d) Filtration Behaviours of Hollow Fiber Membranes

Sheng Chang, School of Engineering, University of Guelph, Guelph, ON, Canada

Hollow fiber membranes play an important role in the water and wastewater treatment. Pressurised and immersed hollow fiber membrane modules are the dominated membrane modules used for the water purification, while for the wastewater treatment, most of the large scale MBR plants use immersed hollow fiber membrane modules in combination with biological wastewater treatment processes. Understanding the filtration behaviour of the hollow fiber membranes is of crucial importance for the membrane module design and the optimisation of the filtration operation condition. This paper presents modeling, experimental, and XMI observation results on the filtration behaviours of the hollow fiber membranes. The modelling results revealed the effect of the fiber dimension, solution and particle properties, and operation conditions on the cake resistance distribution, the non-uniformity of the local flux, and the suction pressure-time profiles. It has been found that the suction pressure time profile, which is usually used as an indication of the membrane fouling under the constant flux condition, could be significantly affected by the fiber diameter, flux level, and size of particles in the solution. The experimental results indicated that under the crossflow or bubbling condition, the small diameter fiber and the axial fiber arrangement could be more beneficial than the larger diameter fiber and the horizontal fiber arrangement, respectively. Through the DOTM observation used in a filtration of a yeast suspension under crossflow and controlled flux conditions, it was observed that some yeast particles randomly deposited and sticked to some specific locations of the surface of a hollow fiber membrane and the attached particles could grow to a large particle conglomeration through constantly attachment of the individual particles from the bulk. The selective deposition to a certain location, instead of forming a cake layer by layer, may imply that the slow irreversible fouling development under the low flux and crossflow conditions could be related to the heterogeneous properties of the membrane and particle surface chemistry. The non-invasive XMI observation revealed the fiber micro-surface properties could have significant effect on the lumen flow of a hollow fiber membrane, resulting in a strong fluctuation in the filtration behaviour of the hollow fibers. Based on the modeling and experimental results, the key factor affecting the filtration performance of the hollow fiber membrane modules will also be discussed.

See more of this Session: Session in Honor of Professor Anthony Fane: Membrane Separations for Sustainable Water and the Environment
See more of this Group/Topical: Topical 1: Separation Needs for Energy Independence and Environmental Sustainability

BF-MBR: PROS AND CONS OF THE POTENTIAL FOR NEW REACTOR DEIGNS AND OPERATION

Prof. TorOve Leiknes,
Department of Hydraulic and Environmental Engineering,
NTNU - Norwegian University of Science and Technology, Trondheim, Norway

Abstract

Coupling biofilm reactors with membrane processes as biofilm membrane bioreactors (BF-MBR) is an interesting development as an alternative technology to activated sludge membrane bioreactors (AS-MBR). Biofilm technology for wastewater treatment can provide substantially lower suspended solid environments for membrane filtration compared to activated sludge processes. Potential benefits are; reduced energy consumption for aeration systems (lower viscosity), less membrane clogging problems, lower fouling potentials, ease of membrane cleaning, and new membrane module designs (i.e. higher membrane packing densities). This study was aimed to propose and investigate alternative membrane reactor designs and process configurations of the BF-MBR process. Alternative and novel biofilm reactor designs have been investigated, for example; a double-deck aerated moving bed biofilm reactor, two-stage aerobic-anaerobic biofilm reactor. The performance and design of the biofilm reactors was aimed to determine TSS control, effluent quality, and enhanced membrane filterability. The biofilm effluent on average had low TSS (typically <50 mg/L), low turbidities and FCOD. Alternative membrane modules and membrane filtration unit designs have been investigated. These include flat sheet systems, tubular and capillary membranes supplied from various manufacturers.

Introduction

In recent years development and commercialization of membrane bioreactors (MBR) for advanced treatment of wastewater has seen an exponential growth. MBRs are commonly understood as the combination of membrane filtration and biological treatment using activated sludge (AS), where the membrane primarily serves to replace the clarifier in the wastewater treatment system. Market growths have been in excess of 10% per year and have essentially been driven by the municipal sector, in particular by the construction of the larger plants with installed capacities above 5.000 m^3/d [1]. The MBR technology is now a cost competitive option for municipal and industrial wastewater treatment, where the applications require exceptional specifications such as enhanced water quality (for bathing water, water reuse), reduced footprint or where the technology is used to upgrade existing plants [2,3].

Although much R&D has been conducted, key bottlenecks of the technology are still being investigated; gaining a better fundamental understanding of membrane fouling and fouling control, reducing the energy demand and consumption, and reducing allover costs through improved operations and increase in system lifetime. Recent research has therefore looked at alternative integrated membrane based wastewater treatment options which potentially can resolve some the challenges found in AS-MBR systems [4].

Coupling biofilm reactors with membrane filtration as biofilm membrane bioreactors (BF-MBR) is an interesting alternative technology to AS-MBRs. Biofilm technology for wastewater treatment can provide a substantially lower suspended solids environment for membrane filtration compared to activated sludge processes. Potential benefits are; less membrane clogging / sludging problems, lower fouling potentials, ease of membrane cleaning, reduced energy consumption for air-scouring [4,5,6]. These characteristics of the BF-MBR also represent an opportunity to design and operate the membrane unit for enhanced solids removal and thus improved membrane fouling mitigation and control as well potentials for using less energy. A tool in achieving this is implementing new membrane module/reactor designs [7].

In recent developments in BF-MBRs, alternative and novel biofilm reactor designs have been investigated. Different biofilm reactor combinations and concepts have been tested to investigate the effect of the biodegradation stage on the performance of the membrane reactor. Different membrane modules have been tested in these studies (*i.e.* flat sheet, capillary / tubular membranes) with varying reactor geometries and designs, and a broad range of operating conditions (*i.e.* fluxes, backwashing settings, air-scouring rates etc.) [7,8,9,10].

Methods

Lab-scale pilot plant trials have been conducted coupling a biofilm reactor based on the moving-bed-biofilm reactor process. The membrane filtration unit was designed as an externally submerged configuration where modifications to the membrane reactor were investigated. A GE Zenon membrane module was installed in the different membrane reactor configurations investigated. Simple models based on classical mass balances of MLSS have been developed and tested to predict the behaviour of the reactor designs [7]. The principle of the pilot plant configuration used and three alternative membrane reactor designs tested are illustrated in Figure 1.

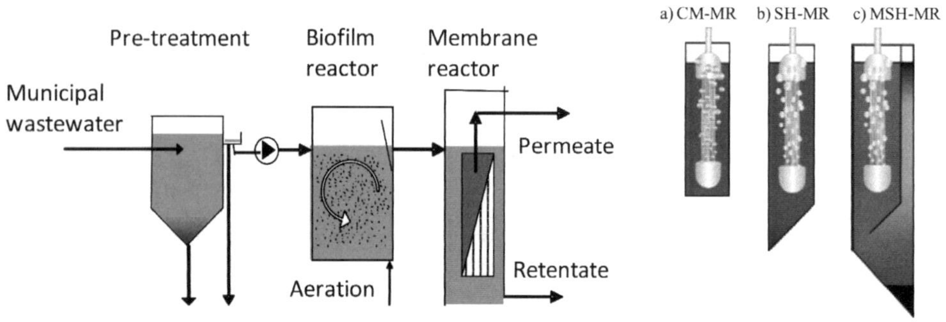

A) Pilot plant configuration and setup B) Alternative reactor designs

Figure 1. Schematic of pilot plant set up (A) and example of alternative membrane reactor designs tested (B) [7]

Three different process configurations and designs of BF-MBR systems have also been investigated [10,11]. Both side-stream module configuration and external-submerged module configurations were tested. The three schemes investigated include: 1) a double-deck aerated biofilm reactor with external flat-sheet membrane filtration unit (membranes supplied by A3 Water Solutions GmbH, Germany); 2) an aerobic moving biofilm-anoxic biofilm MBR with energy recovery from aeration and side-stream tubular membrane filtration unit (Puron membranes supplied by Koch Membrane Systems, Germany); 3) a two-stage aerobic moving biofilm-anaerobic activated sludge MBR with a side-stream tubular membrane filtration unit (membranes supplied by Polymem, France). The principle pilot plant configurations of the systems tested are illustrated in Figure 2.

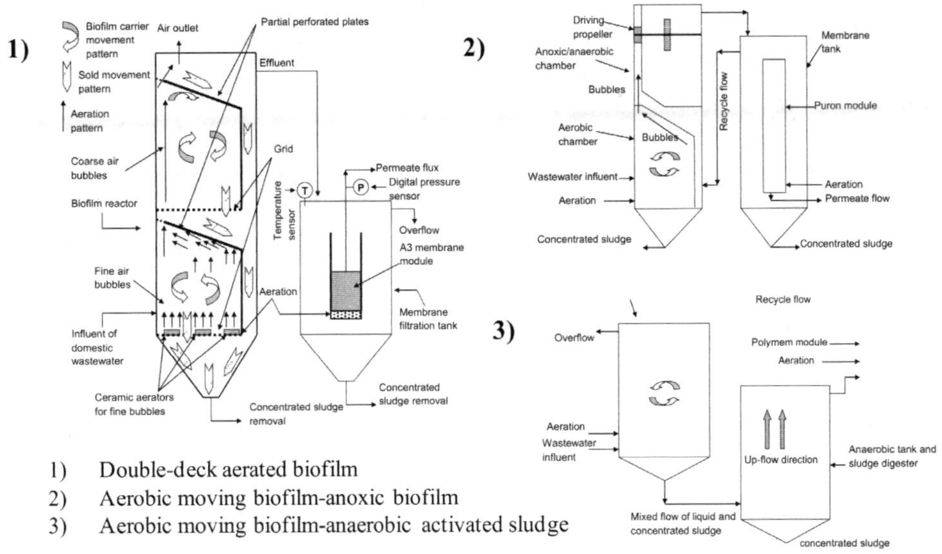

Figure 2. Illustration of alternative BF-MBR process configurations investigated [11]

All experiments were done with municipal wastewater from a combined sewer system which was pretreated using a small gravity settler and then pumped from an overflow to the respective pilot plants. The pilot plants were equipped with National Instruments / LabVEIW data acquisition units and online measurements using various sensors, *i.e.* temperature, pressure, flow etc. All the pilot plants have been operated with varying conditions, *i.e.* flux, specific aeration demand, organic loads etc. Standard water quality sampling and analyses have been conducted, including more detailed studies of the suspended solids and particle size distribution etc. Membrane filtration performance has been assessed by evaluating overall fouling rates as a function of operating conditions.

Results and discussion

The impact of the membrane reactor design was investigated by studying three membrane reactor designs: 1) a completely mixed reactor (CM-MR), 2) membrane reactor with integrated sludge-hopper (SH-MR), and 3) a membrane reactor with a modified sludge-hopper (MSH-MR) [7]. The biofilm reactor was operated with a HRT of 4 hours. Suspended solids (SS) in the effluent from the biofilm reactor varied between 90-150 mg L^{-1}, while COD and FCOD were between 120-242 and 25.9-43.1 mg O_2 L^{-1} respectively. The membrane reactor was operated in a 5 minute cycle with production flux 35 LMH, backwash flux 42 LMH and recovery 96%. Continuous air scouring of the membrane was employed for all tests with specific aeration demand (SAD_m) set on approximately 1 $Nm^3_{air}m^{-2}_{membrane\ area}h^{-1}$.

Steady state concentrations of MLSS around the membrane area for the 3 designs were ~3900, ~1000, ~400 mg/L, giving fouling rates within production cycle of 20, 6, and 3×10^{-5} bar sec^{-1}, respectively. Lower concentrations of MLSS and COD around the membranes as a function of the modified rector designs resulted in a better membrane performance (Figure 3 A)). However, a reduction in MLSS is not directly proportional to a reduction of fouling rates (*i.e.* dTMP/dt). The characteristics of the suspended matter around the membrane also play an important role in membrane fouling, though a reduction of these foulants by an enhanced membrane reactor design is a significant contribution to controlling and minimizing fouling of the membrane. Soluble matter (*i.e.* FCOD) was reduced by ~ 40% in the MSH-MR compared to the CM-MR (Figure 3 B)). The difference in membrane performance for the three alternative designs and selected quality parameters is shown in Figure 3. Particle size distribution analyzed in membrane reactors showed that the differential number percentage of submicron particles around the membrane in the reactors with a sludge-hopper design (SH-MR and MSH-MR) could be reduced by ~ 8 % and ~ 10 % respectively, compared to a completely mixed design (CM-MR). Modification of the membrane reactor in the BF-MBR process was demonstrated to be beneficial and relatively easily done by introducing an integrated flocculation zone with a sedimentation zone beneath membrane unit [5].

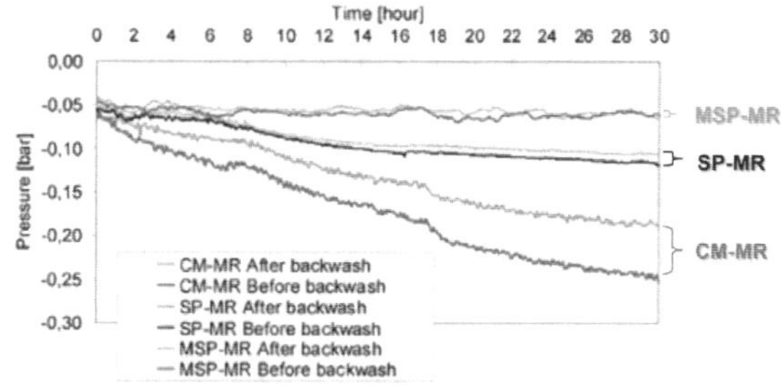

A) Membrane performance

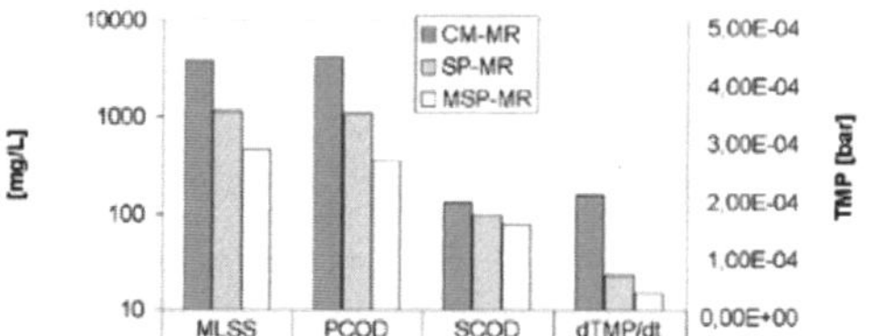

	MLSS (mg/L)	PCOD (mg/L)	SCOD (mg/L)	dTMP/dt (bar/cycle)
CM-MR	3900	4100	131	2,00E-04
SP-MR	1152	1100	98	6,00E-05
MSP-MR	462	353,6	78,4	3,00E-05

B) Process performance

Figure 3. Example of membrane performance and reduction of foulants as function of alter native membrane reactor designs [11]

One the benefits of biofilm processes is that it can provide a substantially lower suspended solids environment for membrane filtration compared to activated sludge processes. Due to the significance of the characteristics and amount of the solids formed in the treatment train, a development strategy of the BF-MBR process was therefore defining alternative process configurations (i.e. both the biological stage and the membrane filtration stage) to maximize and improve solids removal from the overall process. Three different schemes were investigated including: 1) a double-deck aerated biofilm reactor with external flat-sheet membrane filtration unit (membranes supplied by A3 Water Solutions GmbH, Germany); 2) an aerobic moving biofilm-anoxic biofilm MBR with energy recovery from aeration and side-stream tubular membrane filtration unit (Puron membranes supplied by Koch Membrane Systems, Germany); 3) a two-stage aerobic moving biofilm-anaerobic activated sludge MBR with a side-stream tubular membrane filtration unit (membranes supplied by Polymem, France). Illustrated in Figure 3 [11,12].

Depending on the design of the biological reactor, the suspended solids in the inflow to the membrane reactor can potentially be reduced. In the one alternative the biofilm reactor was based on a double-deck approach with two biofilm chambers, each confined by a partially perforated sloping plate on the top of the chambers and a grid plate at the bottom. Fine bubble aeration was used in the lower chamber. The internal walls inside the biofilm reactor created the pathway for biomass particle settling. The concentrated sludge settled at the bottom of the bioreactor and the membrane filtration tank, and removed periodically. With this design the biofilm effluent had less than 45 mg/L SS, resulting in the concentration around the membrane during operation to accumulate to only around 150 mg/L. Based on this performance a new modified membrane module using flat-sheet membranes was designed and tested [8]. Overall a more compact membrane module design was possible, without sludging or module clogging being observed, and resulting in lower membrane fouling rates and less energy needed for air scouring to maintain a sustainable filtration [8].

For the second alternative a process based on a vertical biofilm reactor consisting of two chambers was investigated. The aim of this process configuration design was to 1) reduce the membrane clogging problem from high TSS, 2) reduce viscosity in the system, 3) utilize the excess aeration energy to get an energy optimization, 4) to potentially reduce the membrane fouling. The lower chamber is used for an aerobic digestion and the upper chamber is for anoxic/anaerobic digestion of the organic substances. The design of this configuration is aimed to make use of the excess energy in air flow used in the aerobic chamber to provide some mixing in the anoxic chamber. A bubble-driven propeller is proposed for this design. A recycle flow from the membrane tank to the bottom of the biofilm reactor is used to control the concentrations of the residual organics or colloids around the membrane. The excess sludge is trapped at the bottom of the vertical biofilm and membrane tanks and removed from the system, typically with a concentration of approximately 15 g/L [12]. The impact of aeration intensity on membrane fouling and overall performance was investigated. The system was operated with permeate flux of 20.95 L/m^2.h and aeration rates of 9.5 and 19 L/min. By doubling the aeration rate, the fouling rate was reduced approximately by 45%. (Figure 4)

In the third alternative the process ass composed of an aerobic moving-bed biofilm reactor and anaerobic sludge reactor. The aim of this design was to minimize the operating cost and to optimize the energy consumption for the MBR processes. Wastewater influent was pumped to the aerobic tank and then fed from the bottom of the aerobic tank to the anaerobic tank. The supernatant from the anaerobic tank was then pumped to membrane filtration unit. No sludge was removed from the aerobic tank as both settled sludge and MLSS were pumped to the anaerobic tank. The sludge waste from the aerobic tank is thus anaerobically digested and the amount of the sludge waste is thus significantly reduced. Solids concentration of the waste removed from the anaerobic tank was approximately 26 g/L. The process alternative showed overall good treatment efficiencies, operated with relatively low aeration demand while achieving low membrane fouling rates.

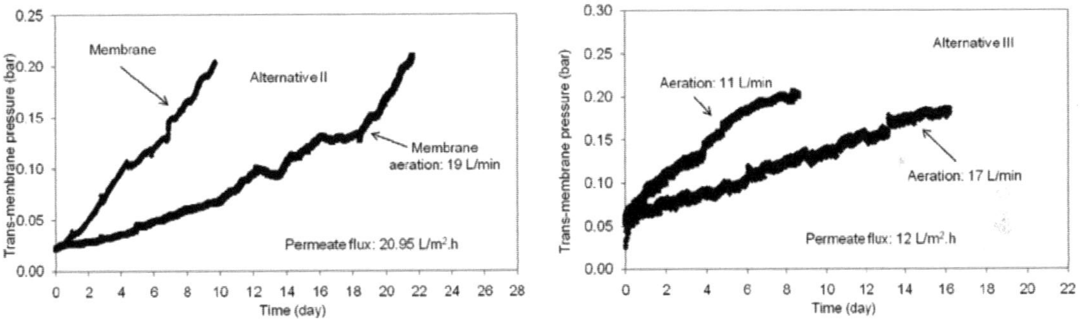

Figure 4. Impact of aeration intensity; process alternative 2 and 3 respectively [12]

One of the goals of the alternative process configurations proposed was as to develop a process concept for suspended solids control, energy optimization method, and sludge waste management. For the alternative designs studied, management of the suspended solids was achieved resulting in low average TSS concentrations in the inflow to the membrane filtration unit. Combined with integrated sludge management strategies a further reduction in the volume of excess sludge removal from the process was also achieved. This had further a beneficial impact on the overall membrane performance by eliminating sludging/clogging problems and reducing both membrane aeration demands and fouling.

Conclusions

High quality effluents were achieved independent of the operating conditions, though the performance of the membrane filtration unit is affected by the operating conditions of the system. In general, the BF-MBR showed that stable operation of the membrane unit with low fouling rates, overall less energy demands, and sustainable operation can be achieved. By coupling a biofilm process with membrane separation there is a potential to maximize the biodegradation stage as well as the membrane filtration stage.

Modification of the membrane reactor in a BF-MBR process is beneficial for establishing stable operation, where the introduction of an integrated flocculation zone with a sedimentation zone beneath membrane unit is one strategy. Such a modified rector design provides a significantly lower concentration of MLSS and COD around the membranes and thus a better overall membrane performance. The proposed simplified model and equations for calculating steady state values of MLSS inside the membrane reactor have been verified, however, a refinement of the model with the aim of developing a more general design tool for improved membrane reactor designs for the BF-MBR process still needs to be investigated.

The potentials of biofilm-MBR processes are that new and flexible process configurations are possible, they offer an alternative strategy for solids control and management which results in lower suspended solids load on the membrane. Furthermore, the operating conditions achieved may provide enhanced membrane performance / less fouling and overall lower energy requirements. Although there is room for optimization and improvement of the treatment schemes, the BF-MBR configurations with alternative membrane reactor unit designs investigated in these studies are found to be competitive treatment schemes for treating municipal wastewater.

Acknowledgements

The experimental work and results presented in this paper have primarily been conducted by PhD candidate Igor Ivanovic (NTNU), and Postdoc Fellow Jirachote Phattaranwik (NTNU). Financial support by: Norwegian research council (NFR) and EUROMBRA research project supported by the European Commission under the Sixth Framework Programme, Contract No. 018480; Krüger Kaldnes, Norway, for support with the biofilm reactor and ZENON Environmental Inc., Canada, for supplying the membrane modules.

References

1. Lesjean, B., Huisjes, E.H. (2008), "Survey of the European MBR market: trends and perspectives". *Desalination,* 231 (1-3), 71-81.
2. C. Brepols, A. Janot, K. Drensla, H. Schäfer, N. Engelhardt (2008) "Considerations on design and financial feasibility of large scale membrane bioreactors", *Conference Design and operation of membrane plants for water, wastewater, industrial effluents*, Amsterdam. October 2008
3. C. Brepols, H. Schäfer, N. Engelhardt (2009) "Economic aspects of large scale membrane bioreactors". *Final MBR-Network Workshop "Salient outcomes of the European R&D projects on MBR technology"* 31 March – 1 April, Berlin 2009 (Germany) ISBN 978-3-9811684-5-7
4. T.O. Leiknes and H. Ødegaard (2007) "The development of a biofilm membrane bioreactor", *Desalination*, 202 (2007) 135-143
5. I. Ivanovic, T.O. Leiknes and H. Ødegaard (2008) "Fouling control by reduction of submicron particles in a BF-MBR with an integrated flocculation zone in the membrane reactor". *Separation Science and Technology*, 43(7) (2008) 1871-1883
6. I. Ivanovic and T.O. Leiknes (2008) "Impact of aeration rates on particle colloidal fraction in the biofilm membrane bioreactor (BF-MBR)". *Desalination*, 231 (2008) 182-190
7. Igor Ivanovic and TorOve Leiknes (2010) "Membrane reactor design as a tool for better membrane performance in a biofilm MBR (BF-MBR)", *Desalination and Water Treatment,* In Press

8. J. Phattaranawik and T.O. Leiknes (2009), “Double-deck aerated biofilm membrane bioreactor with sludge control for municipal wastewater treatment”, *AIChE Journal*, 55 (2009) 1291-1297
9. J. Phattaranawik J. and T.O. Leiknes (2010), “Study of hybrid vertical anaerobic sludge-aerobic biofilm membrane bioreactor for wastewater treatment”, *Water Environment Research*, 82 (3) (2010) 273-280.
10. Leiknes, TorOve; Phattaranawik, Jirachote; Ivanovic, Igor (2009). “Challenges and potentials of biofilm-MBR for municipal wastewater treatment”, *Final MBR-Network Workshop "Salient outcomes of the European R&D projects on MBR technology" Book of Proceedings*. Berlin, Germany, 2009, ISBN 9783981168457
11. TorOve Leiknes, Jirachote Phataranawik, Igor Ivanovic (2009), “Prospects and potentials of biofilm-MBRs for municipal wastewater treatment”, in *WEFTEC 2009 Conference,* October 2009, Orlando, Florida. Proceedings.
12. Phattaranawik, J., Leiknes, T. (2009), “Investigation of alternative process configurations for biofilm-MBR systems for municipal wastewater treatment”, *Proceedings, WISA-MTD 2009 Conference*, Stellenbosch, South Africa, 12-15 May 2009

(357a) All Solid State Rechargeable Lithium-Air Batteries Using Block Copolymers

Daniel T. Hallinan Jr. and Nitash P. Balsara, Department of Chemical Engineering, University of California, Berkeley, Berkeley, CA

The growing need for alternative energy and increased demand for mobile technology require higher density energy storage. Existing battery technologies, such as lithium ion, are limited by theoretical energy density as well as safety issues. Other battery chemistries, such as lithium-air, are promising options for dramatically increasing energy density. Safety can be improved by replacing the flammable, reactive liquids used in existing lithium-ion battery electrolytes with polymer electrolytes. Block copolymers are uniquely suited for this task because ionic conductivity and mechanical strength, both important properties in battery formulation, can be independently controlled. In this study, lithium-air batteries were assembled using lithium metal as negative electrode, polystyrene-b-poly(ethylene oxide) copolymer with lithium salt as electrolyte, and a porous positive electrode. The positive electrode consisted of polymer electrolyte for ion conduction, carbon for electron conduction, and manganese dioxide as catalyst. Batteries were charged and discharged over many cycles. The battery cycling results were compared to a more conventional battery chemistry, lithium iron phosphate. To our knowledge, this is the first demonstration of an all solid state rechargeable lithium-air battery.

See more of this Session: Polymers for Energy Production and Storage
See more of this Group/Topical: Materials Engineering and Sciences Division

(357b) Imidazolium-Based Polymerized Ionic Liquids for Solid Polymer Electrolytes: The Effect of Anion Type

Yuesheng Ye and Yossef A. Elabd, Chemical and Biological Engineering, Drexel University, Philadelphia, PA

Solid polymer electrolytes (SPEs) with high ionic conductivities are of great interest for energy conversion and storage devices (e.g., fuel cells, batteries, solar cells, actuators, capacitors), in part, because they overcome shortcomings of liquid electrolytes (e.g., leakage, flammability, toxicity, stability). The utilization of ionic liquids (ILs) in SPEs has gained remarkable interest due to the excellent physiochemical properties of ILs (e.g., negligible vapor pressure, nonflammability, high ionic conductivity, good electrochemical, chemical and thermal stability). Recently, polymerized ionic liquids (PILs), a new class of SPEs, have received increased attention for energy applications.

This study focuses on the synthesis and characterization of imidazolium-based PILs. In this work, an imidazolium-based IL monomer, 1-[(2-methacryloyloxy)ethyl]-3-butylimidazolium bromide (MEBIm-Br), was synthesized and subsequently polymerized using conventional free radical polymerization. To investigate the effect of anion type on polymer properties and ion conduction, the polymer precursor, poly(MEBIm-Br), was exchanged with other anions such as tetrafluoroborate (BF4), bis (trifluoromethanesulfonyl) imide (TFSI), trifluoromethanesulfonate (Tf), and hexafluorophosphate (PF6). Also, a novel method was developed to prepare the imidazolium-based PIL with hydroxyl anions. The type of anion was shown to have a significant effect on the glass transition temperature of the polymer and subsequently the ionic conductivity. Additionally, the anion type also had a significant effect on the hydrophilicity of the polymer, where the addition of water to hydrophilic systems resulted in significant increases in ion transport. Detailed results of anion effect on the ion transport will be presented.

See more of this Session: Polymers for Energy Production and Storage
See more of this Group/Topical: Materials Engineering and Sciences Division

(357c) Nanotube Forests for Electrochemical Energy Storage From Electrostatic Assembly

Lin Shao[1], Woo-Sik Jang[2] and **Jodie Lutkenhaus**[1], (1)Department of Chemical Engineering, Yale University, New Haven, CT, (2)Department of Chemical Engineering, Yale Universtiy, New Haven, CT

Electrical energy storage is increasingly important as the global need for efficient, carbon-free, sustainable energy production rises. Whether the energy comes from renewable or non-renewable sources, an efficient means of storage is needed for responsible energy consumption. Currently, lithium-ion batteries are prevalent in many of these applications because of their established reliability and superior performance relative to older technologies (i.e. Ni-Cad batteries); however, for energy storage at increasingly smaller dimensions, breakthroughs in electrode design are required. Nanostructured electrodes are poised to replace conventional two-dimensional planar electrodes and address many issues (i.e., surface area, diffusion distance, energy and power density).

Here, we combine layer-by-layer (LbL) assembly and nanotemplating to realize LbL-nanotube array cathodes containing vanadium pentoxide and polyaniline. V2O5 stores and intercalates Li+ ions and polyaniline conducts electrons while acting as a binder. Layer-by-layer assembly is performed on porous membranes, the membranes are selectively removed, and nanotube forests are left behind. The goal is to create a high surface area electrode that minimizes the diffusion resistance of Li+, which could boost energy and power density. The growth behavior of the V2O5/polyaniline assembly is monitored using UV-Vis spectroscopy and scanning force microscopy (SFM). Electrochemical properties are characterized using cyclic voltammetery. Scanning electron microscopy (SEM) and transmission electron microscopy (TEM) images confirm that large areas of LbL nanotubes can be made in this fashion. Future work will assess how these nanostructured cathodes will behave electrochemically as nanotube aspect ratio is varied.

See more of this Session: Polymers for Energy Production and Storage
See more of this Group/Topical: Materials Engineering and Sciences Division

(357d) Molecular Dynamics Simulations of Ionic Aggregates in a Coarse-Grained Ionomer Melt

Lisa M. Hall, Mark J. Stevens and Amalie L. Frischknecht, Sandia National Laboratories, Albuquerque, NM

Ionomers--polymers containing a small fraction of covalently bound ionic groups--have potential application as solid electrolytes in batteries. Understanding ion transport in ionomers is essential for such applications. Due to strong electrostatic interactions in these materials, the ions form aggregates, tending to slow counterion diffusion. A key question is how ionomer properties affect ionic aggregation and counterion dynamics on a molecular level. Recent experimental advances have allowed synthesis and extensive characterization of ionomers with a precise, constant spacing of charged groups, making them ideal for controlled measurement and more direct comparison with molecular simulation.

We have used coarse-grained molecular dynamics to simulate such ionomers with regularly spaced charged beads. The charged beads are placed either in the polymer backbone or as pendants on the backbone. The polymers, along with the counterions, are simulated at melt densities. The ionic aggregate structure was determined as a function of the dielectric constant, spacing of the charged beads on the polymer, and the sizes of the charged beads and counterions. The pendant ion architecture can yield qualitatively different aggregate structures from those of the linear polymers. For small pendant ions, roughly spherical aggregates have been found above the glass transition temperature. The implications of these aggregates for ion diffusion will be discussed.

*Sandia National Laboratories is a multi-program laboratory operated by Sandia Corporation, a wholly owned subsidiary of Lockheed Martin Corporation, for the U.S. Department of Energy's National Nuclear Security Administration under contract DE-AC04-94AL85000.

See more of this Session: Polymers for Energy Production and Storage
See more of this Group/Topical: Materials Engineering and Sciences Division

(357e) Anion Exchange Membrane Fuel Cells

Hyea Kim, Murat Unlu, Junfeng Zhou, Irene Anestis-Richard and Paul A. Kohl, Georgia Institute of Technology, Atlanta, GA

Proton exchange membrane fuel cells (PEMFCs) face many obstacles including the high cost of platinum catalyst and perfluorinated membranes, high electro-osmotic drag of fuel, and slow kinetics for the electrochemical reactions. Anion exchange membrane fuel cells (AEMFCs) have the potential to address many of the problems facing proton-based cells. The high pH environment in AEMFCs provides faster kinetics for both anode and cathode reactions and paves the way for use of non-noble catalysts, such as silver and nickel. The use of hydrocarbon membranes may lower the cost of cells.

In an effort to develop an AEMFC, new types of anion exchange membrane (AEM) have been synthesized, including epoxy cross-linked AEMs. The conductivity and stability of the membranes has been investigated in terms of ion exchange capacity, and solubility. The ionomer, which is a soluble form of a membrane in the catalyst, plays an important role in achieving high performance. In this work, the formulation of the AEM electrode has been investigated and integrated with different types of AEMs in an effort to achieving high catalytic activity.

Moreover, new fuel cell designs have been conceived using alkaline membranes. A hybrid fuel cell, which consists of proton exchange membrane and AEM electrodes, has an advantage at both electrodes due to the more facile reaction kinetics at high pH. In addition, a new fuel cell stack design has been described using an AEMFC and PEMFC in series operating with a single fuel tank servicing both anodes. The advances in membrane technology and cell design will be discussed.

See more of this Session: Polymers for Energy Production and Storage
See more of this Group/Topical: Materials Engineering and Sciences Division

(357f) Molecular Simulations of Neat, Hydrated, and Acid-Doped Polybenzimidazoles

Shuo Li and J. R. Fried, University of Cincinnati, Cincinnati, OH

Polybenzimidazoles are a set of linear aromatic polymers that exhibit excellent mechanical strength, thermal stability, and high resistance to acids, bases, and oxidative attack. These properties have led to their use in a variety of applications from fire-resistant materials to membrane applications. In particular, acid-doped polybenzimidazoles have shown promise as proton-exchange membranes (PEMs) for fuel cell use up to 200 °C. The most frequently used dopant is phosphoric acid (PA), introduced in 1995 by Wainright. As generally observed, acid doping level and water uptake are two critical factors that influence not only the conductivity but also the mechanical strength of the membranes. Ma et al. also proposed a comprehensive conductivity mechanism for proton transfer (PT) through the polymer-acid-water system based on their experimental results, and noted that hydrogen bonding played an important role in the organization of solvent molecules and the polymer chains to promote conductivity.

While there have been many experimental studies of membrane conductivity and structural and mechanical properties available in the literature, simulation studies of the acid-doped polybenzimidazole system have not been reported. In this study, molecular dynamics simulations have been used to investigate the density, X-ray diffraction patterns, radial distribution functions (RDFs), diffusion properties, and, most importantly, the hydrogen bonding behavior of the neat, hydrated, and PA-doped benzimidazole polymer systems under different levels of water uptake, acid doping, and temperature. Results are compared experimental data available in the literature. Poly(2,2'-*m*-phenylene-5,5'-bibenzimidazole) (PBI, Fig. 1A), poly(2,5-benzimidazole) (ABPBI, Fig. 1B) and poly(*p*-phenylene benzobisimidazole) (PBDI, Fig. 1C) are three important examples of benzimidazole polymers. PBI, with a reported glass transition temperature, T_g, of 420 °C, is the only commercially used polybenzimidazoles and the majority of research has been focused on the acid-doped PBI membranes. ABPBI, which has the simplest repeat-unit structure in the family, has also been widely investigated for applications as PEMs. Although PBDI has not been widely studied as potential PEMs, its hydrated structure has been characterized through X-ray crystallography of a similar diimidazole model compound. The polymer-acid-water interactions illustrated in that study provided valuable insight into the behavior of solvent molecules for both the hydrated and acid-doped polymer systems.

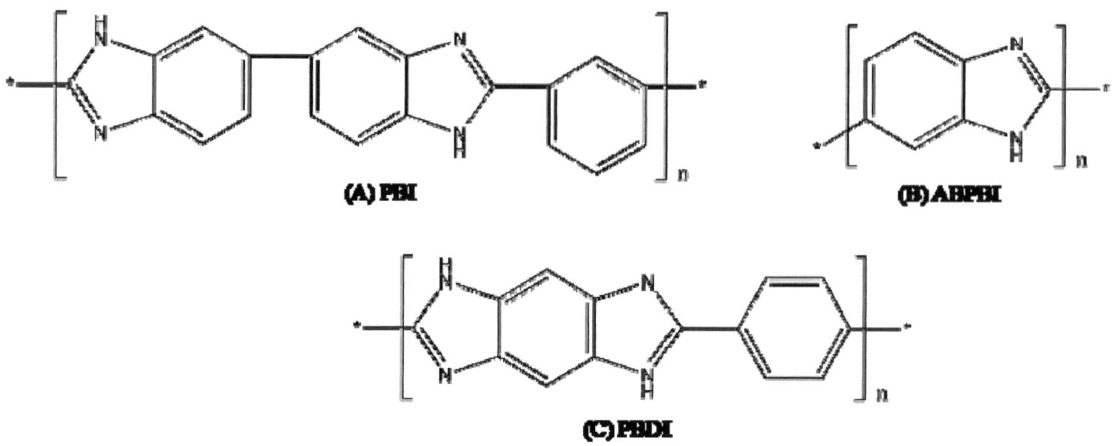

Figure 1 Structures of poly(2,2'-*m*-phenylene-5,5'-bibenzimidazole) (PBI), poly(2,5-benzimidazole) (ABPBI), and poly(*p*-phenylene benzobisimidazole) (PBDI)

As a first step in this study, density, X-ray diffraction patterns, RDFs, and hydrogen bonding behavior were analyzed for all there neat benzimidazole polymers and their corresponding hydrated and PA-doped systems at the end of 800-ps NPT dynamics. Single chains of each of the three benzimidazole polymers were built head-to-tail at random torsion angles to represent the neat polymers. The chain length was 75 repeat units (RUs), 200 RUs, and 100 RUs for PBI, ABPBI, and PBDI, respectively. The hydrated systems were constructed by mixing the polymer chains with water molecules to make the ratio of imidazole units to water molecules to be 1:2. Similarly, the acid-doped system was constructed by mixing polymer chains with both PA and water molecules to make the ratio of imidazole units to PA molecules to water molecules to be 1:2:2.

Second, the effect of the number of polybenzimidazole chains used in the simulation, the water uptake, acid doping level, and temperatures on structural and diffusion properties, and evidence of hydrogen bonding behavior in the neat, hydrated, and PA-doped ABPBI systems were studied. Three 80-RU ABPBI chains were built at 25 °C and 180 °C, respectively, to represent the neat ABPBI. Three 80-RU ABPBI chains and different numbers of water molecules (the ratio of imidazole units to water molecules was 1:0.5, 1:1, 1:1.5, 1:2, 1:2.5, and 1:3, respectively) were constructed at 25 °C to represent the hydrated systems of interest. For the acid-doped systems, two 80-RU ABPBI chains were selected to represent ABPBI. At 25 °C, two acid-doped systems were constructed by mixing ABPBI chains with both PA and water molecules to make the ratio of imidazole units to PA molecules to water molecules to be 1:1.5:1.25 and 1:3:2.5, respectively. At 180 °C, another two acid-doped systems were built. One of them was constructed by mixing ABPBI chains with both PA and water molecules to make the ratio of imidazole units to PA molecules to water molecules to be 1:3:2.5. The other was constructed by mixing ABPBI chains with PA molecules only to make the ratio of imidazole units to PA molecules to be 1:3. Density, X-ray diffraction patterns, RDFs, and hydrogen bonding behavior were analyzed and compared between the constructed systems at the end of the 800-ps NPT dynamics. For the four acid-doped systems, another 3-ns NVT dynamics was run at 298 K or 453 K, following the NPT dynamics, to calculate the self-diffusion coefficients.

Good agreement between available experimental data and the simulated density, X-ray diffraction patterns, RDFs, and self-diffusion coefficients of the neat, hydrated, and acid-doped polybenzimidazoles provides validation for the simulation results. The results, especially the hydrogen bonding behavior, are consistent with the expected PT mechanism in polybenzimidazole-based PEMs.

(428a) A Multi-Effect Membrane Distillation Process for Preconcentration of Aqueous Solutions of Sugar in the Production of Bioethanol

Kun Yao[1], Yingjie Qin[2], Liqiang Liu[3], Yingjin Yuan[1] and Dandan Liu[3], (1)Department of Biochemical Engineering, School of Chemical Engineering and Technology, Tianjin University, Tianjin, China, (2) Chembrane Research & Engineering,Inc., Newark, NJ, (3)Chembrane Engineering & Technology, Inc., Tianjin, China

Bioethanol production is generally regarded as a promising alternative way to substitute the traditional petroleum-based liquid fuels. Lignocelluloses, which are abundant in nature and independent of the market for food and cattle feed, are preferable to conventional starch or sugar containing feedstocks as raw materials for large scale production. The reported concentration of fermentable sugars in the cellulosic hydrolysate prior to fermentation is relatively low because of some process restraints during hydrolysis to obtain high sugar yield. However, in order to decrease the size of footprint of the entire processing plant and also energy consumption in downstream separation/purification, the feed to fermentation requires high concentration of sugars. To resolve this contradiction, an economically efficient and well performed preconcentration unit is necessary to added between the hydrolysis and fermentation step.

In the present study, multi-effect membrane distillation (MEMD), a new membrane based distillation process, has been developed, which combines the advantages of both membrane distillation (MD) and multistage flash (MSF) by equipping air gap membrane distillation (AGMD) with internal heat recovery. A novel separation device in the form of hollow fibers is fabricated to test its separation performance, which is identified as MEMD module. The water vapor flux (J) and energy efficiency in term of performance ratio (PR) and thermal efficiency (η) are the most important indicators for evaluation of module performance. J indicates the productivity of the membrane module; PR tells how much energy is recovered by internal configuration and η shows how much energy is lost due to conduction according to the second law of thermodynamics. Experiments were conducted using the dilute salt aqueous solution as a tracer to investigate the influences of operating variables including inlet temperatures of two sets of different fibers (Th:70-90°C and Tc: 25-45Å°C) and flow rate (F:16-48L/h) on these above three performance parameters. During the experiment, no leakage is detected and the distillate is of good quality which means the separation efficiency is almost 100%. It is found that the most important determinant parameter is flow rate and there exists trade-off phenomenon between flux (J) and energy efficiency (PR and η) under experimental ranges, that is, the maximum flux will be obtained with high temperature Th, low temperature Tc and high flow rate F while the maximum PR or η is obtained with high temperatures Th and Tc, as well as low flow rate F. Two kind of modules with different hollow fiber diameter and length (M1 and M2 fabricated in Chembrane, China) are also compared. Response surface method (RSM) is carried out to build an empirical quadratic model for prediction of the separation performance and optimization, which agrees well with the experimental results. In general, the typical measured water permeation flux in this study is around 2.0 - 9.0L/m2 h and the value of PR is 4 - 12 and the value of η is 0.80 - 0.95. Flux decline together with reduction of PR and the decrease of η were observed with the increased sugar concentration (up to about 500g/L) mainly because of its high viscosity and there existed slightly differences among three sugars (glucose, xylose and sucrose). But it should be noted that the performance of this new device is still appreciated even at higher concentration. This MEMD process was successfully applied in preconcentrating sucrose aqueous solution from 150g/L to 600g/L and also obtains 14-fold of initial concentration of model cellulosic hydrolysate (mixture of glucose and xylose aqueous solution) in a very efficient way.

See more of this Session: Separations for Renewable and Sustainable Energy Development I
See more of this Group/Topical: Topical 1: Separation Needs for Energy Independence and Environmental Sustainability

(428b) Removal of Char Particles From Bio-Oil by Liquid-Phase Microfiltration

Asad Javaid[1], Xiaoming Pan[1], Surita R. Bhatia[2], George W. Huber[3] and **David M. Ford**[1], (1)Chemical Engineering, University of Massachusetts, Amherst, Amherst, MA, (2)Chemical Engineering, University of Massachusetts Amherst, Amherst, MA, (3)Chemical Engineering, University of Massachusetts-Amherst, Amherst, MA

Bio-oil formed by the fast pyrolysis of biomass has tremendous potential as a renewable feedstock to make liquid transportation fuels. However, the presence of char particles in the bio-oil causes problems in storage and end-use. Currently there is no well-established technology to remove char particles less than 10 microns in size. In this study we describe the application of a liquid-phase microfiltration (MF) process to remove char particles from bio-oil down to slightly sub-micron levels. Tubular ceramic membranes of nominal pore sizes 0.5 and 0.8 microns were employed to carry out the MF, which was conducted in the cross-flow mode at temperatures near 40 degrees Celsius and at different trans-membrane pressures varying from 1 to 3 bar. Microscopic and ash content analysis of the feed and permeate streams were conducted to determine the efficacy of the process. Our results demonstrated the removal of a large percentage of the char particles with a significant reduction in overall ash content of the bio-oil. Other than the removal of the suspended chars, the chemical composition of the liquid bio-oil was changed very little by the MF process. Results of fouling analysis obtained from longer runs of bio-oil through the membranes clearly showed that the cake formation mechanism of fouling is predominant in this process. A membrane cleaning protocol for the fouled membrane was also developed.

See more of this Session: Separations for Renewable and Sustainable Energy Development I
See more of this Group/Topical: Topical 1: Separation Needs for Energy Independence and Environmental Sustainability

SOLID–LIQUID SEPARATIONS IN PILOT- AND COMMERCIAL- SCALE OPERATIONS

Dennis Gertenbach[1], and Brian L. Cooper[2]

Paper 428c, for presentation at the AIChE National Meeting, November 10, 2010, Salt Lake City, UT

ABSTRACT

Benchtop processes that are developed with the intent of being run at the demonstration scale present a multitude of technical challenges. Solid-liquid separations are among the most common process operations that become large and often unanticipated capital and operational costs due to insufficient development effort in the research phase. Often separation technologies from established industries can be applied to new processes to mitigate the risk involved with scale up. The presentation will focus on Hazen's experience with solid-liquid separations, including dissolved air flotation (DAF), counter current decantation (CCD), as well as various incarnations of filtration and centrifugation.

SOLID-LIQUID SEPARATIONS

Separation costs are difficult to isolate given the complexity of the chemical industry but have been estimated to account for between 40 and 70% of capital and operating costs in a commercial process.[3] The principles and technologies are well-established and criteria for equipment selection are available, yet these unit operations continue to present difficulties in scale up and operations. Separation processes are often not given sufficient attention at the pilot scale; it is more common to adopt a trial-and-error method in development until a piece of equipment that offers promise is selected at the bench scale. Very often, a small variance in the process can render the specified equipment useless and cause a return to the design phase or worse, development at scale.

All mechanical separation equipment, with the possible exception of those used for impingement, are based on the principles of sedimentation or filtration,[4] as shown in Figure 1.

---

[1] Senior Vice President, Hazen Research, Inc., gertenbachd@hazenresearch.com
[2] Project Manager, Hazen Research, Inc., cooperb@hazenresearch.com
[3]Spear, M., 2009. Separations in flux. http://www.chemicalprocessing.com/articles/2006/025.html (accessed September 27, 2010).
[4]Pierson, H.G.W., "The Selection of Solid-Liquid Separation Equipment", Solid-Liquid Separation, L. Svarovski, 2nd ed., 1981, p. 525.

Figure 1. Sedimentation and filtration schematics[5]

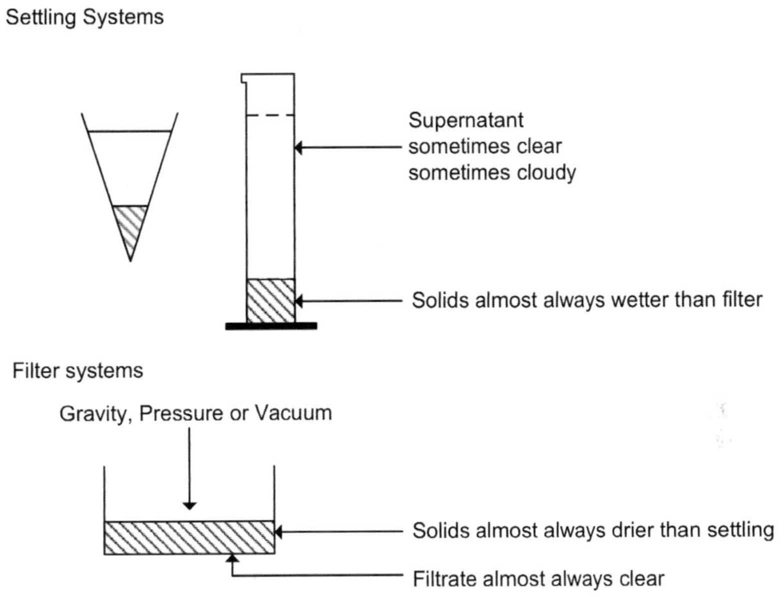

Each technique has numerous configurations and applications, along with well-established strengths and weaknesses, yet all are dependent on the nature of the two phases in the slurry. Often, separation unit operations that have not had sufficient attention in the research phase of a process will fall to centrifugation or belt filtration equipment. Both of these technologies allow for a measure of flexibility (which makes them a relatively safe bet to accomplish a separation), but the respective capital costs of these technologies are very high. Continuous vacuum filtration equipment costs generally range from $1,400 to $3,700/ft^2 FOB point of manufacture. Scroll-conveyor sedimentation centrifuges are dependent on the required bowl diameter, but can cost up to $150,000 for simple carbon steel. Special materials of construction, high throughputs, and material handling issues that require a large unit such as a Sharples Super D-Canter can run as high as $300,000.[6] Installed costs will vary depending on the equipment requirements, but for commercial centrifuge units, infrastructure and ancillary equipment can generally be expected to be 1–3 times the purchase price.[7]

---

[5]Ibid., p. 526.

[6]Peters, Max S., Timmerhaus, Klaus D. and West, Ronald E., Plant Design and Economics for Chemical Engineers, 5th ed. McGraw-Hill, 2003, pp. 864-867, corrected by Annual Index (2003-2009) Chemical Engineering Plant Cost Index (CEPCI), Chemical Engineering, Volume 117, No. 7, July 2010, p. 60.

[7]Perry, Robert H., and Green, Don W., editors, Perry's Chemical Engineer's Handbook, 7th ed., McGraw-Hill, 1997, pp. 18-125.

Savings are certainly realized when the development step is minimized, but short-cutting the generation of proper data to select and properly size the solid–liquid separation equipment usually results in poor performance. Some of the most expensive capital equipment in a process requires larger-than-expected operating costs, and can potentially become the rate-limiting step in an entire process, necessitating an expensive retrofit.

A number of excellent publications have covered the topic of solid–liquid separation technologies and equipment selection criteria. The current paper will attempt to introduce the core principles found in these texts, comment on our experience with these technologies, and provide a brief overview of the process options available to the engineer. For further reference, a complete bibliography has been included in the last section of the document. In general, the examples follow the Duty Specification and Separation Characteristics classification first published by Purchas and Wakeman in 1986.[6]

The first step in tackling the solid–liquid separation needs for a process is to define the requirements of the separation. Aside from the general scale of the proposed operation, the product phases must be defined; if both the solids and liquids are to be collected, the choice of available technologies may be somewhat limited. Generally, a solid–liquid separation is used to recover 1) valuable solids such as crystals formed in a purification process, 2) a valuable liquid such as a plant extract, 3) both phases such as leachate and tails both containing mineral values, or 4) neither phase such as the removal of particulates of a wastewater stream prior to discharge. Other considerations are the acceptable yields, required throughput, the value of the products(s), potential feedstock variation, the ability to introduce reagents to aid in the separation, the requirement for batch integrity and cleanliness, and the acceptable liquid content of the solids, both before and after a potential wash step. Engineering concerns include the footprint of various technologies, as well as required infrastructure, capital costs, lead times, and maintenance costs and labor.

It quickly becomes clear that specifying a particular unit operation for solid–liquid separations is complex and presents a significant challenge in most cases. It is crucial that experimentation at the laboratory scale for a process include studies on slurry behavior and sedimentation/filtration tests. One advantage of a thorough assessment of available technologies is the revelation that multiple techniques may be employed to effect the required balance between efficiency and throughput. In this way, what normally appears as a single operation in the classical process flow diagram becomes a small process itself. This is shown in Figure 2.

Figure 2. Solid–liquid separation as a multi-component process[8]

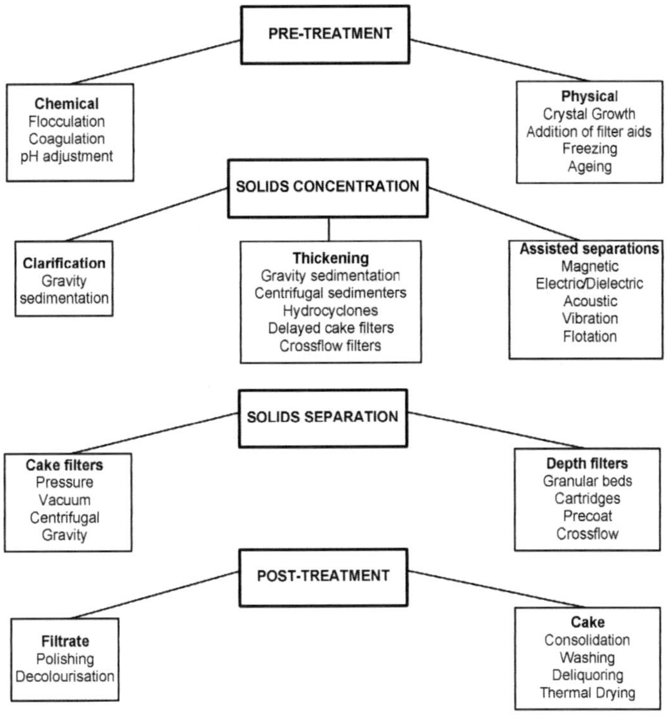

If one considers the possibility of pretreating or concentrating a particularly large process stream, separation and post-treatment options may still be required, but the slurry stream will in theory be much smaller, opening the door to more capital-intensive technologies that would previously have been untenable. A number of the options listed in the chart are covered in this document, along with some general notes on predicting behaviors and laboratory programs that can test the suitability of a given system.

It should be noted that chemical treatment of a slurry is a common and often very effective practice that enables the use of many of the techniques described in this document. However, the subject is not within the scope of the current document. The principles and techniques for chemical treatment of slurries (addition of coagulants and flocculants, filter aids, etc.) is covered in depth in many of the sources listed in the bibliography.

SEDIMENTATION

Sedimentation is based on the behavior of solids suspended in a fluid under the influence of a number of forces, often the force of gravity, but also potentially

[8]Wakeman, R. and Tarleton, S., Solid/Liquid Separation: Principles of Industrial Filtration, Elsevier Advanced Technology, 2005, p. 2.

centrifugal, buoyant, and compression forces. The particle flow through a fluid is described by the Reynolds number,[9] which is essentially a ratio of the inertial forces (in the simplest case, gravity or centrifugal force) to viscous forces (drag force and buoyancy).

$$Re=\rho ux/\mu$$

Where ρ and μ are the density and viscosity of the fluid, respectively, u is the particle-fluid relative velocity, and x is the particle size. In systems for which the particles are very small or nearly neutrally buoyant or for which the fluid is sufficiently viscous, the Reynolds number will be small; for most solid–liquid systems that present separation difficulties, the solids are present as fine particles and are difficult to separate, and Re < 0.2. This is important, as the flow around the particles as they settle is laminar and makes possible the use of the Navier-Stokes equations for calculating the drag force, and allows one to assume that the time to the terminal settling velocity can be assumed to be close to zero. The calculation for determining the terminal settling velocity then becomes:[10]

$$u_t = x^2 (\rho_s-\rho)g/18\mu$$

This parameter is key in evaluating potential technologies that could be used in the separation process. In Hazen's experience, values above 0.4–0.5 m/h warrant experimental verification and further study to size equipment, whereas values below this range indicate that the size of the equipment required to effect a gravity sedimentation will be prohibitively expensive and that gravity separation is not a feasible solution.

The calculated terminal settling velocity only represents a general idea of how a slurry will perform in sedimentation. As the solids settle out of solution, their behavior changes due to their increasing proximity to other particles in solution, to the point where several zones of concentration are formed (see Figure 3). Eventually, a point is reached at which the sedimentation presents a single interface between the solids and the clear liquid. This is known as the critical separation point; the sedimentation behavior of the solids is fundamentally changed, and the settling velocity no longer represents the movement of the particles; the sedimented pulp slowly compresses as liquid is forced upwards through channels in the deposit.

---

[9] Ibid., p. 20.
[10] Ibid., p. 22.

Figure 3. Plots of interface height and solids concentration versus time for batch settling[11]

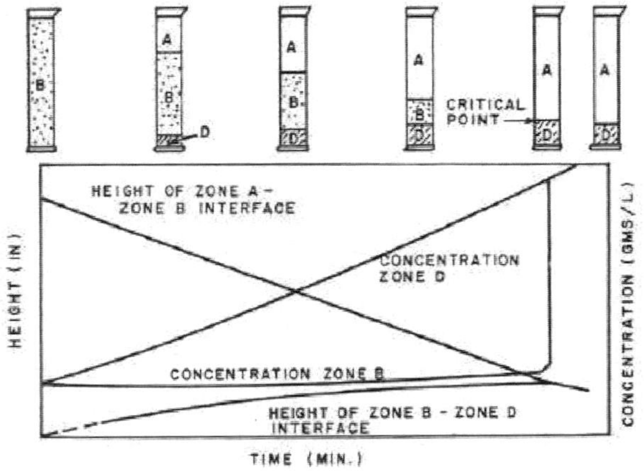

The calculated settling behavior of particles in a slurry can be verified by a simple Kynch test, in which a slurry of known volume is allowed to settle in a graduated cylinder. By measuring the height of the interface over time and sampling the final settled solids to determine their concentration, a clear picture of the slurry behavior will usually indicate if the expected settling rate is borne out and, more importantly, how appropriate gravity sedimentation may be for a particular process. Utilizing a plot of the data generated during the settling test, the Kynch method will also allow for a calculation of the required unit area to sediment a slurry in ft^2/short ton/day (or m^2/metric ton/day), which will drive design decisions and the subsequent cost/benefit analysis. The equation used to generate this figure includes a correction for the depth of the settled solids based on empirical comparisons between full-scale thickeners and lab-scale cylinders.[12]

TYPES OF SEDIMENTATION

Sedimentation is generally divided into two types of processes. Thickening refers to the concentration of solids in a feed stream, while clarifying refers to the removal of solids from a relatively dilute feed stream.[13] Thickeners/clarifiers operate on the same principles and are, in their most basic form, circular settling tanks with a rake to move the settled solids to a central underflow discharge. To aid in the separation, the bottom of the tank is generally sloped towards a discharge port in the center. These vessels are often large, as this allows for sufficient throughput given the typical terminal velocity of settled solids. The

[11]Figure from Cheremisinoff, Nicholas, Pocket Handbook for Solid-Liquid Separations: Calculations and Guidelines for Process Engineering, Gulf Publishing Company, 1984, p. 89.
[12]For details, see Seidel, D. C. "Laboratory Procedures for Hydrometallurgical-Processing and Waste-Management Programs", United States Department of the Interior Bureau of Mines Information Circular 9431, 1995.
[13]Cheremisinoff, p. 85.

overflow and the underflow can be collected continuously, allowing for an extremely large process flow given the correct conditions. A cross section showing the various parts of a basic thickener is shown in Figure 4.

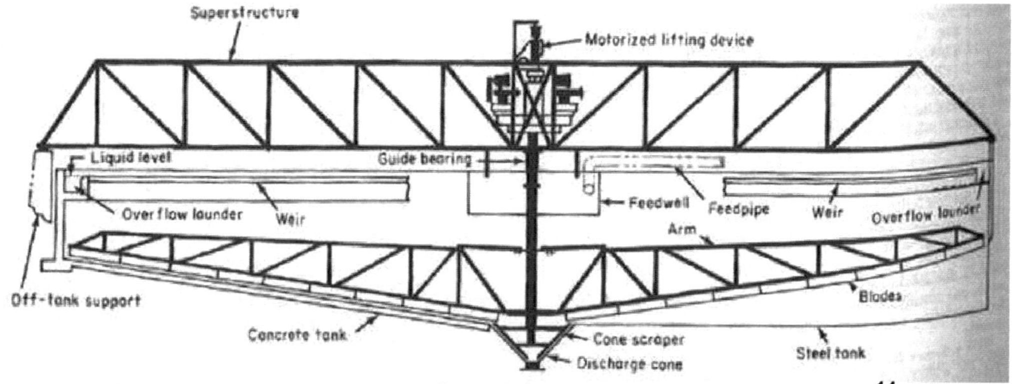

Figure 4. Unit thickener (EIMCO Process Equipment Co.)[14]

A very common use of sedimentation technology is with continuous countercurrent decantation (CCD). This application includes a number of thickeners arranged in series, where the feed enters the circuit in one direction and the underflow is continually pumped upstream, resulting in a bulk clarified liquid flow in one direction (the same as the feed), and the bulk flow of concentrated solids slurries moves in the other direction. A diagram of a CCD process is shown in Figure 5.

Figure 5. Schematic diagram of a CCD process

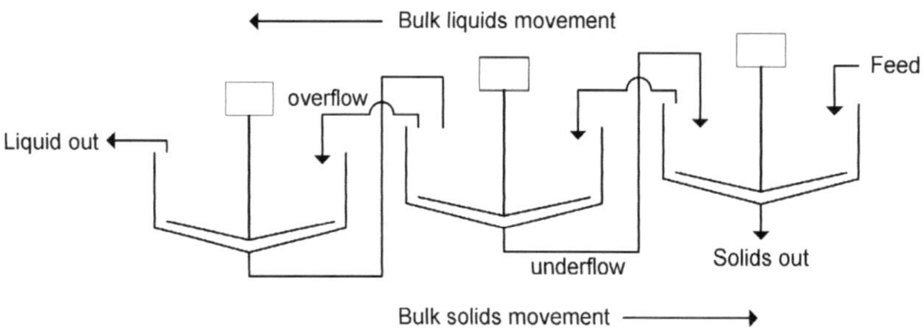

One of the advantages of this type of operation is the ability to tolerate large swings in the feed composition. Because the holdup volume of these types of systems is so large, short-term slurry concentration variances are absorbed without much difficulty.

[14]Perry and Green, pp. 18-64.

FLOTATION

Flotation processes rely on the affinity of the solids for the solution in which they are contained and are generally applied to aqueous systems in which the introduction of a gas will alter the buoyancy of the solids through adhesion and subsequent gas–solid agglomeration. Many factors (nature of the solids, size of the gas bubbles formed, surface tension, etc.) affect the behavior of a flotation system, and extensive experimental programs are often needed to determine the optimal setup. Flotation is governed by the same equation as sedimentation, the difference being that now the relative density of the solids will be less than that of the liquid and will rise in solution, eventually at the terminal velocity described by:

$$u_t = x^2 (\rho_l - \rho_g) g / 18\mu$$

Where, in this case, x^2 is the diameter of the gas–solid agglomerate, and ρ_l and ρ_g are the respective densities of the liquid and the gas.[15] This equation is applicable for solutions with low Reynolds numbers (< 1) and gas bubbles that are sufficiently small (~0.2 mm).

In general, the smaller the bubbles the better the flotation, because smaller particle sizes will also aid in the adhesion process. Generally, it is easier to alter the size of the bubbles in a separation system, as the particle sizes are often driven by preceding steps in the process that created the slurry. For systems in which high mixing rates are possible (the solid phase is not friable), dispersed air flotation can be used, where the gas in entrained in the solution by vigorous mixing. A more common technique is dissolved air flotation (DAF), where the gas and the slurry are mixed under pressure, which is then released, causing the formation of very small gas bubbles. An example of a test setup for this is shown in Figure 6.

Figure 6. A DAF test apparatus[16]

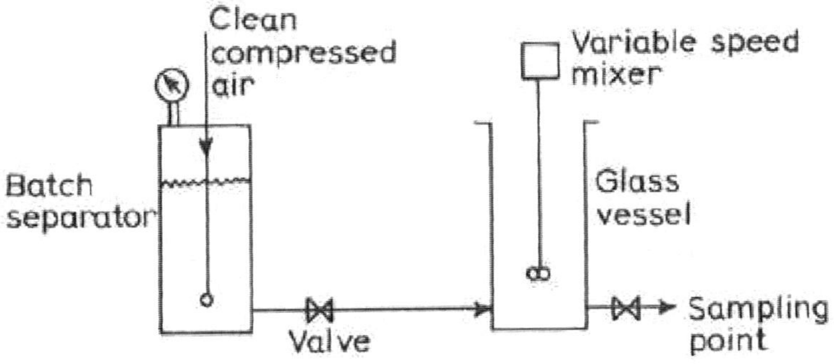

[15]Matis, K.A. and Zouboulis, A.I., "The Role of Bubble/Particle Size", Flotation Science and Engineering, Marcel Dekker, Inc., 1995, p. 79.
[16]Gochin, R.J., "Flotation", Solid-Liquid Separation, 2nd ed., edited by Svarovsky, Butterworths and Co., 1981, p. 517.

A number of different flotation technologies have been developed with varying geometries and applicability. Separation efficiency, sensitivity to system inputs (gas introduction rate location and methods, necessity for positive bias (air bubble/solids agglomerates retain their buoyancy) through the net downflow of wash water), selectivity, and capital requirements have led to the development of robust simple configurations, as well as very specialized and fully engineered flotation systems, such as the Jameson Cell, shown in Figure 7.[17]

Figure 7. Layout of a Jameson Cell

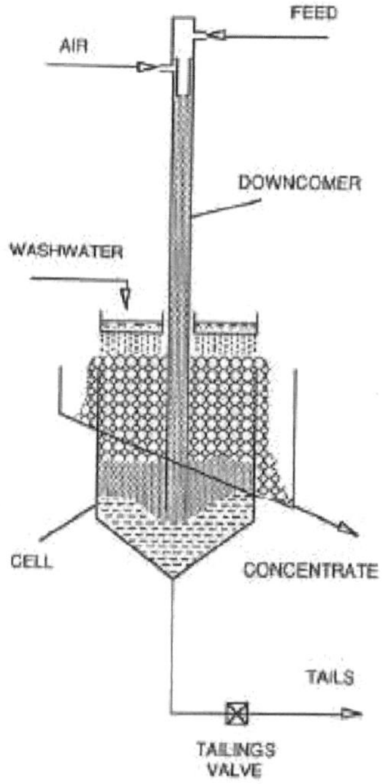

The Jameson Cell allows for significant contact time between the gas and the feed in the downcomer. Once the mixture moves into the cell, the reduction in the downward superficial velocity allows the particle/bubble agglomerates to rise to the surface as a froth, which then overflows the cell and is collected.[18]

HYDROCYCLONES

The hydrocyclone is a cono-cylindrical vessel with an outlet at the bottom and at the top, fed by a tangential inlet as shown in Figure 8.

[17]Evans, Geoffrey M., Atkinson, Bruce W. and Jameson, Graeme J., "The Jameson Cell", Flotation Science and Engineering, Marcel Dekker, Inc., 1995, p. 334.
[18]Ibid, p. 333-334.

Figure 8. Basic hydrocyclone diagram[19]

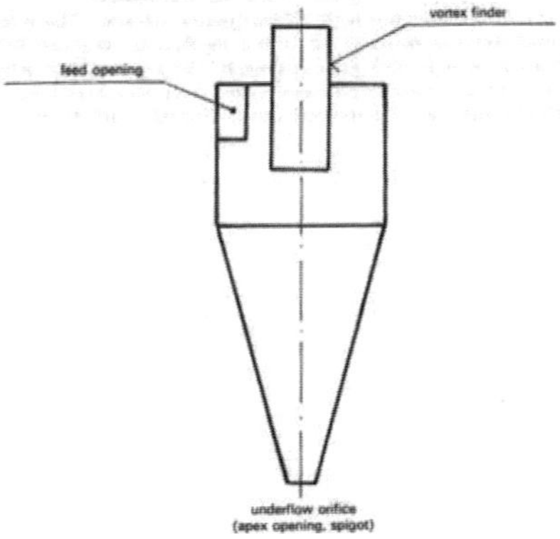

The separation of solids in the hydrocyclone is typically size-dependent, and the terminal velocity of the particle can be described by a slightly more complicated expression of the Stokes equation:

$$u_t=(\Delta\rho/\rho)\tau^*a$$

Where $\Delta\rho$ is the density difference ρ_s-ρ, τ^* is the particle relaxation time $\rho_s x^2/18\mu$, and a is the field acceleration of particles, which changes with particle position.[20] It should be noted that the primary strength of a hydrocyclone separation is in its ability to make fine cuts in particle size, once properly sized. Rather than a wholesale solids separation, this technology is best utilized as a pretreatment/solids conditioning step, aiding the efficiency of a subsequent solids separation step. A good example of this would be fines removal using a hydrocyclone that would allow for better flow through a filter bed (see p. 13).

Key to the operation of the hydrocyclone is the sizing and geometric arrangement of the cone, as well as the outlet downcomer, commonly called the vortex finder. Once a stable vortex is formed in the hydrocyclone body, the presence of the vortex finder will create circulation eddies that generate areas of lower resistance where fine particles can flow from the high-pressure areas near the inner cyclone wall towards the low-pressure areas in the center of the unit. A central air core usually forms in the center of the hydrocyclone, an indication of a stable vortex and a place where dispersed and even dissolved gases will report during operation.[21]

Particle movement in a slurry flowing in a hydrocyclone can be described by the Reynolds number according to the following equation:

---

[19]Svarovsky, L., Hydrocyclones, Hold, Rinehart and Winston Ltd., 1984, p. 30.
[20]Ibid., p. 7.
[21]Ibid., p. 33.

$$Re = (v \cdot D \cdot \rho)/\mu$$

Where v is the characteristic velocity (tangential component of the fluid velocity), D is the hydrocyclone diameter, and ρ and μ are again the density and the viscosity of the liquid, respectively. Using this general behavioral equation, one sees that an increase in diameter will result in a larger Reynolds number and hence a rougher cut; finer particles with a slight density difference from the liquid phase will be more inclined to report to the underflow. For this reason, hydrocyclone applications are often designed to include multiple units working in concert as opposed to a single large unit. Fortunately, this also aids in the hydrocyclone behavior, as very large units must account for gravity's influence, while smaller units can be installed in just about any orientation as needed.[22]

CENTRIFUGAL SEDIMENTATION

Centrifugal sedimentation also relies on the density difference between liquids and solids subjected to a centrifugal force. Operation of the centrifuge allows for the application of much greater forces than are possible in a hydrocyclone, enabling more efficient separations. Assuming the concentration of solids in a slurry is sufficiently low that interaction between the particles can be neglected, the behavior of solids in a sedimenting centrifuge can be described by a further modification of Stokes' law:

$$dr/dt = (\Delta\rho x^2 r\omega^2)/18\mu$$

Where r is the radial position of the particle of size x, t is the time interval that the particle is subjected to the centrifugal acceleration, $\Delta\rho$ is the density difference between the solid and the liquid, ω is the angular velocity, and μ is the fluid viscosity.[23]

Due to the complex nature of the interactions between slurry components and the body of a centrifuge, numerous factors must be considered when scaling a sedimenting centrifuge, particularly when evaluating units that are not of similar geometry. One way of attempting to evaluate this is to consider the Sigma factor, which can normally be approximated as:

$$\Sigma \approx (\omega^2/g)\pi L(^3/_2 r_3^2 + {}^1/_2\, r_1^2)$$

Where ω is the angular velocity, L represents the full length of the settling zone, and r_3 and r_1 represent the total radius of the rotating portion of the centrifuge and the distance between the center axis and the start of the settling zone, respectively. This is shown in Figure 9 below.

---

[22]Ibid., p. 83.

[23]Svarovsky, L. "Separation by Centrifugal Sedimentation", Solid-Liquid Separation, 2nd ed., edited by Svarovsky, Butterworths and Co., 1981, p. 192.

Figure 9. Schematic diagram of a simple tubular centrifuge[24]

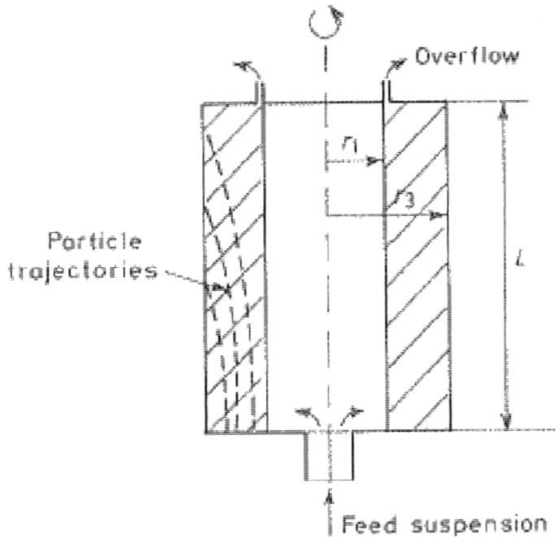

The Sigma factor is a widely used criterion for the comparison of centrifuges of similar geometry, liquid flow patterns, and the same centrifugal gravity (*G*-force)[25] and allows for the approximation of unit performance at scale from experimental data taken with a smaller unit. The Sigma factor makes some assumptions that cause the correlation to be inexact; Stokes law is likely not entirely descriptive of the settling solids behavior over the entire range of *G*, as the solids in the liquid stream may contain entrained liquid once settled, eddies and flow disruptions probably occur at the entrance and the exit of the unit that are disregarded, etc. To offset these known weaknesses in the Sigma factor, efficiency factors are utilized, which are often well understood and constant (tubular centrifuges) or can vary widely (decanting centrifuges) depending on the nature of the settled solids removed and other considerations.

Great care must be taken in selecting a centrifuge for a commercial operation as the capital expense, as well as operational and maintenance costs for these unit operations, can be very large.

Sedimentation technologies are very useful in solid–liquid separations and are often powerful yet inexpensive techniques that can be used to concentrate solids before the final solid–liquid separation. Likewise, if a particular particle size in a slurry is the desired product, the use of sedimentation principles can often achieve the product separation. The major disadvantage of the majority of these technologies is the requirement for continuous operation, precluding any ability to maintain batch integrity, as well as the addition of large amounts of liquid in order to achieve the solids concentration desired. While this is not an issue for many industries, those that are more heavily regulated from a quality standpoint may find this problematic.

---

[24]Ibid., p. 191.
[25]Perry and Green, pp. 18-116.

FILTRATION

The fundamental difference between filtration and sedimentation is, of course, the existence of a barrier in filtration that is impermeable to one of the constituents of a slurry. The flow of liquid through the cake that builds on the barrier (filter media) can be described by Darcy's basic filtration equation:[26]

$$Q = K(A\Delta p/\mu L)$$

Where Q is the flow through the bed of solids, K is the permeability of the bed, A is the surface area of the filtration, Δp is the driving pressure, μ is the fluid viscosity, and L is the thickness of the bed. Naturally, as a filtration progresses, many of these factors are changing and most of them are independent of the equipment used, save for designs that are utilized to benefit the flow (i.e., removing a cake from the filter surface during operations so that the bed thickness remains small). Typically, pretreatment or solids concentration of a slurry will greatly aid filtration, as particle behavior impacts flow to a significant degree. In sedimentation this is not such a concern, as the settling of solids displaces entrained liquid but no actual flow through the settled solids is required.

There are two general classes of mechanical filtration: surface filtration, which works largely by direct interception of the particles from the slurry as it passes through the surface of the filter, and depth filtration, which utilizes a medium that has increasingly dense layers from the upstream to the downstream side. The two types are shown in Figure 10.

Figure 10. A surface filter compared with a graded-depth-type filter.[27]

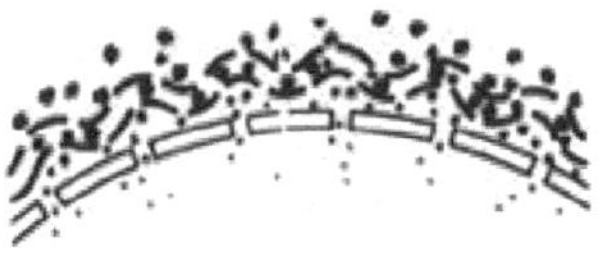

Surface Filter

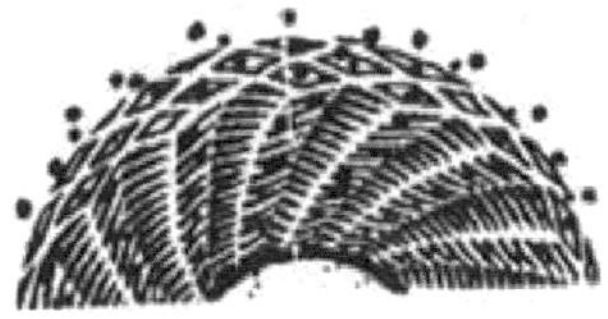

Graded Depth Filter

Applications that utilize surface filters generally encounter a reduction in the effective pore size, as accumulating particles gradually close off pores. This phenomenon can be exploited to aid in a filtration (see below), but can also be the cause of a functioning filtration unit suddenly losing any effective flow through the bed as the pore sizes become so fine that they will not allow sufficient liquid flow. Softer particles that are more prone to deformation under the influence of force can also block pores in a surface filter, which is one of the reasons why

---

[26]Svarovsky, L. "Filtration Fundamentals", Solid-Liquid Separation, 2nd ed., edited by Svarovsky, Butterworths and Co., 1981, p. 244.
[27]Warring, R.H., Filters and Filtration Handbook, 1st ed., Gulf Publishing Company, 1981, pp. 17-18.

filtering gels can present a significant problem. The filter shown in Figure 10 offers a potential advantage in that the fine particles that initially pass through the filter can be re-loaded on the unit after the cake has formed. The larger solids on the filter surface then act as a filtering aid for the finer particles. This is a common occurrence and is often utilized in commercial applications.

Solids that are removed by a depth filter, on the other hand, are subjected to a much longer and more complex path before they are removed entirely from the liquid flow. The depth filter is designed to allow (relatively) constant liquid flow by exposing the solids to a very large surface area that contains varying pore sizes. Because the surface area is so much larger and the effective smallest pore size is so much smaller, a greater pressure drop across the filter is usually required to obtain equivalent flow rates with a surface filter. The strength of the technology is in the depth filter's ability to remove much larger amounts of solids, particularly those that are difficult to remove by surface filtration, all while retaining the ability to collect very fine particles and maintaining reasonable flow rates.

FILTER SELECTION

The selection of filter media is critical to the success of a filtration operation, such that the media warrants a brief overview. As always, the following represents a general overview of the options available to the engineer, and more complete lists can be found in the references utilized in writing this paper. Generally, the media types discussed here range from those that give the roughest form of separation to those that give the most precise. These different categories have significant overlap, in that one type of filter may include a subset of another.

STRAINERS

The simplest filter media is the strainer, which is normally used to offer protection of equipment or process streams from minute amounts of solid contaminants above a specified size. Strainers can take the form of a flat disc or a conical basket and may be constructed from simple perforated plates or supported wire mesh. Perforated plates typically provide coarse filtration for particles down to 150 µm, while wire-mesh strainers can accommodate solids down to about 40 µm, although with a requirement for additional support as the wire mesh can be very fine.[28] In larger pipelines, a basket-style strainer is usually preferred, both for its relative capacity to collect solids and for ease of removal and cleaning. A typical basket strainer is shown in Figure 11.

[28]Ibid., pp. 67-71.

Figure 11. Diagram of a basket-style strainer.

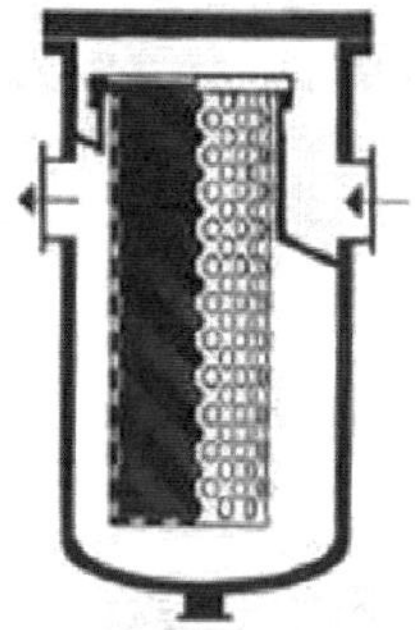

SCREENS

Screens are porous barriers that are designed to capture a particular particle size, while allowing smaller particle sizes to pass through. Strainers would be considered screens that are designed for incidental solid removal; most screens are utilized to accomplish a much more complete particle separation and oftentimes serve as support for other filter media. The type of screen and the size range of the feed to be treated will be the major factors determining the capacity of a screening operation.[29]

Screens can be static (fixed) or moving (reciprocating) depending on the application. The simplest static screen for removing larger particles is the grizzly, which is a set of parallel bars held apart by spacers that create regular openings. A slightly more complex version of this same layout is the wedge wire deck, where the "bars" have a specific geometry that helps to prevent screen blinding (solids plugging all the apertures in the screen and preventing further flow through the screen). An illustration of a wedge wire screen is shown in Figure 12.

Figure 12. Illustration of a wedge wire screen[30]

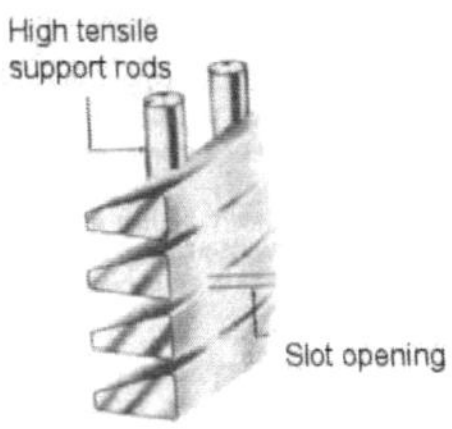

[29]Osborne, D.G. "Screening", Solid-Liquid Separation, 2nd ed., edited by Svarovsky, Butterworths and Co., 1981, pp. 215-217.
[30]Warring, p. 73.

The geometry is crucial in this design, as any solids that manage to make it through the slot opening will then pass freely through the screen. Installing a wedge wire screen in the reverse fashion would almost certainly ensure that the screen will be blinded when separating solids with diameters close to that of the slot opening.

In general, screens are subjected to significant abrasion and must be constructed of sufficiently robust material to function with expected efficiency over time. This is especially true for reciprocating screens that are utilized in filters that introduce movement to aid in separations. Once the desired aperture size and surface area (percentage of open area) have been specified, it must then be determined whether the screen will be woven wire, round or wedged bars, "comb" style (round bars that are not fixed so their rotation will naturally aid in clearing the buildup of solids), or a simple perforated plate. Additionally, some screens are conical, and many screening applications utilize multiple screens with progressively smaller apertures to achieve a high throughput and efficiency.

CLOTH AND PRECOAT/FILTER MEDIA

Filters can utilize "cloth" in the form of woven wire, paper, adsorbent media, and synthetic fibers. The media are often inert solids such as diatomaceous earth that can be loaded onto a screen or other support to provide a porous surface on which solids accumulate; however, chemically active filter media can be used if required by the separation. Typically, the type of cloth or media utilized is that which allows the greatest flow through the bed without creating difficulties in cleaning the removed solids from the media.

MEMBRANES

Filter membranes represent a separate class of media, as they allow for the use of filters that are capable of separating extremely fine particles (removal of bacteria or even viruses) from liquid. Membranes vary in their configuration, but are commonly pleated cartridges, flat sheets, or hollow fibers.[31] Microfiltration (pore size measured in microns) and ultrafiltration (pore size measured in Daltons) are made possible by the use of membranes, which are most commonly made of polysulphone, polyacrylamide, polyvinylidenedifluoride, polyethylene, and ceramics. As one would expect, the pore sizes are very small, so the pressure required to enable flow through the membranes during operation can be quite high.

The type of filter cloth or media required is usually determined in the development of processes that are to be scaled, or at least the nominal particle size of the solids to be separated is known. Often a filtration, whether it utilizes a

[31]Wakeman, R. and Tarleton, S., Solid/Liquid Separation: Scale-up of Industrial Equipment, Elsevier Advanced Technology, 2005, p. 140.

cotton cloth on a belt filter, a woven wire screen, or a synthetic fiber membrane, will perform better as a separation progresses, but experience with this serendipitous phenomenon is generally obtained through multiple runs of a separation at scale and cannot be estimated or counted on in the design phase.

TYPES OF FILTRATION

The simplest type of commercial filtration is a screening operation in which large particles are separated on a mesh that is large enough to catch the particles but provides essentially no resistance to the liquid flow. Typically, these have been used to dewater coarse materials, such as mineral ore and coal. The nature of the particles in these systems (whether the fines contain values or can be lost to the drain, as well as the overall particle size distribution) has a major influence on the screen material selection and design. The driving force for the separation is typically the flow of the liquid through the screen, though some screening operations utilize an additional pressure drop, such as a vacuum. A typical dewatering process is shown in Figure 13.

Figure 13. A dewatering process including a cyclone to pretreat the slurry[32]

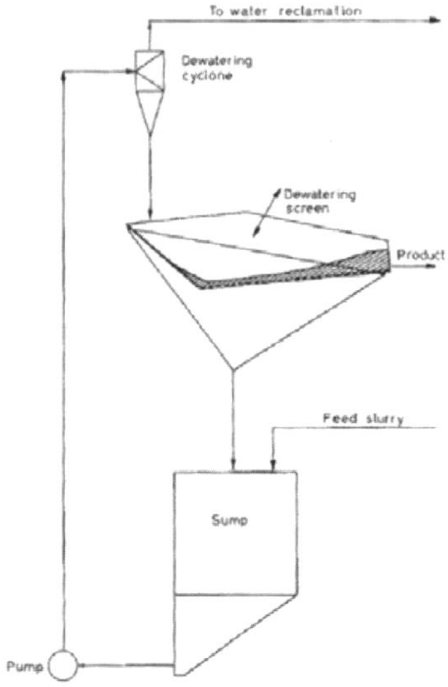

Screening operations are often subjected to abrasive conditions due to the material being separated. Dewatering applications can be quite disparate in their designs, and can utilize a solids concentration step prior to the screen, vibrating screens, deck inclination or curved screening surfaces, a spiraling vortex, vacuum to increase the pressure across the screen, and other advanced equipment, depending on the behavior of the solids and the desired product.

---

[32]Warring, p. 233.

VACUUM FILTRATION

Solid liquid separations can be readily accomplished in many processes utilizing a vacuum filtration, however the major limitation is that the pressure driving the flow through the cake will be somewhat less than 1 atmosphere (14.7 psi). Additionally, the lower pressures involved tend to liberate volatile solvents in a slurry which can present significant challenges at scale. Despite these disadvantages, vacuum filtration is widespread and used in almost all industries that require bulk liquid/solid separations.

The rotary vacuum drum filter is a cloth-covered perforated drum suspended over a trough containing the slurry to be separated. The fundamental design of the rotary drum vacuum filter is relatively unchanged since it was patented by James and William Hart in 1872.[33] Anywhere from 25-75% of the filter area can be submersed in the slurry trough, but most commonly approximately 35% of the filter area is utilized. Slurry is pulled onto the surface of the drum that is submerged and the solids are retained on the cloth. They then rotate with the drum out of the vessel, often where they are washed and then removed from the cloth surface by means of a backflow of compressed air, a scraper, or both. A schematic of the rotary vacuum drum filter is shown in figure 14.

Figure 14. A schematic of a rotary vacuum drum filter[34]

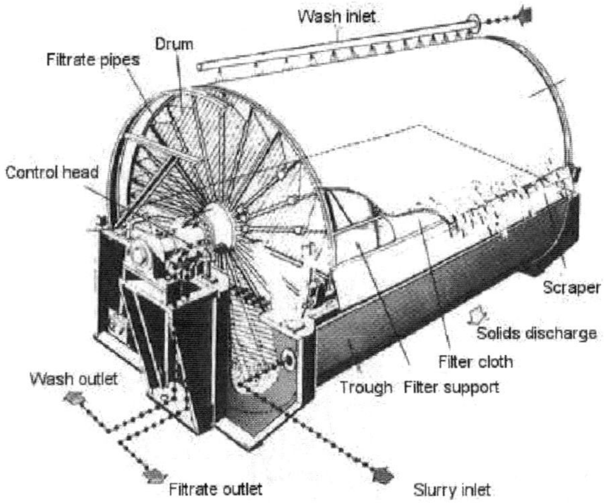

Because the slurry is subject to gravity, it must be mixed while the drum filter is utilized in order to ensure that the solids do not sand out in the trough instead of being removed on the filter. Additionally, only the area of the drum that is not submerged can effectively dewater the cake, so the mechanism used to feed the unit (amount of filter area submerged) is in competition for filter area with the

---

[33] Wakeman, Richard and Tarleton, Steve, Solid-Liquid Separation: Scale-up of Industrial Equipment, Elsevier Inc, 2005, p.244

[34] Warring, p.124.

washing and dewatering portions of the drum. Drum filters can be used in precoat applications, and can utilize cake compression to aid in removal of mother liquor or wash. Top feed drum filters are also used, however the effective filter area of these units is limited by the geometry of the drum and by gravity, and thus not very common except in very specialized applications.

Another example of a commercial vacuum filtration unit is the horizontal vacuum belt filter (HVBF), which like the rotary drum filter allows the filter cloth to be washed and recycled during the operation. In the more complex configurations of this technology, the belt is advanced through the use of a continuously moving belt with a support tray that moves over horizontal vacuum chambers. Breaking the vacuum at the end of a cycle (when the tray must reset by sliding under the belt back to its original position), allows for constant feed of the slurry to the belt, limited only by the speed at which the cake dewaters.[35] A schematic of a HVBF is shown in figure 15.

Figure 15. A schematic of a Pannevis RT/RB horizontal vacuum belt filter

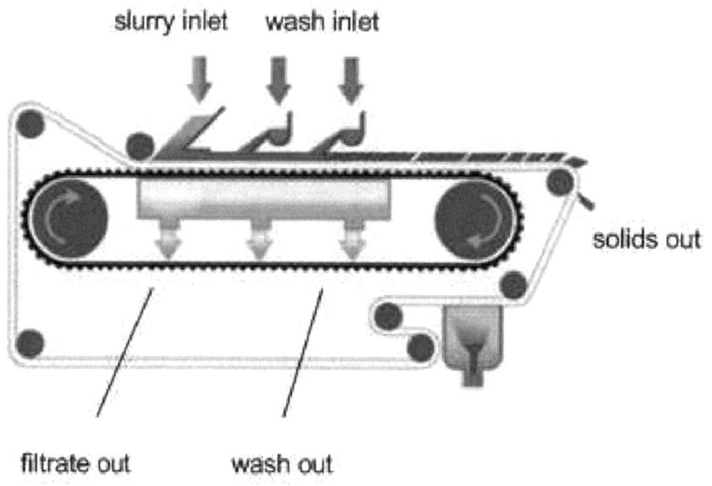

The horizontal vacuum belt filter generally has a very large footprint, and while it operates on similar principles with the drum filter, it does have the advantage that the belt speed can be adjusted to allow for sufficient dwell time to dewater the cake.[36] This capability is not possible on a drum filter, as the feed characteristics will change if the rotation of the drum is simply slowed. The belt filter can slow the filtration while keeping the feed constant, allowing for the exploitation of thicker beds if it suits the process. Additionally the filtrate and the wash can be collected separately as the unit can be run so the different streams report to separate collection vessels. Units like the HVBF have many moving parts, and while generally reliable, they do require regular maintenance.

[35] Wakeman, Richard and Tarleton, Steve, Solid-Liquid Separation: Scale-up of Industrial Equipment, Elsevier Inc, 2005, p.254
[36] Ibid., p. 254.

PRESSURE FILTRATION

The pressure limitations of vacuum filtration can be overcome by using pressure filtration. The larger Δp provides a greater driving force, giving a greater throughput for a given filtration area. The solids cake that is collected is often compressible, such that the cake permeability changes during a pressure filtration, affecting the flow through the bed. To account for this, data that relate to the specific resistance of a cake (α) and the cake porosity (ε) to the pressure differential (Δp) and the effect of particle size are described by:

$$\alpha = (180/\rho_s x^2)\cdot(1-\varepsilon/\varepsilon^3)$$ [37]

Cake resistance (R_c, as opposed to the resistance of the filter itself, R) is defined by the specific resistance (α), multiplied by the area of the filter (A), as:

$$R_c = \alpha A$$

From these, it is understood that the pressure filtration will be generally described by a simplified form of Darcy's Law:

$$Q = \Delta p/(\mu(R_c+R))$$ [38]

In general, the resistance of the cake, R_c, will increase as the cake compresses. This is usually a constant process, which is why most pressure filtration equipment is operated in batch mode, to allow for removal of the cake from the filter media and reduce the cake resistance so the operation can start again. It is inherently more expensive to operate than continuous equipment because of added labor requirements.

Filter presses, which are a series of vertical chambers produced by stacked plates of varying designs to create small regions between each plate, are commonly used to accomplish these separations at scale.[39] A simple diagram of two types of filter presses is shown in Figure 16.

---

[37] Wakeman and Tarleton (Solid/Liquid Separation: Principles of Industrial Filtration), p. 110.
[38] Ibid., p. 101.
[39] Warring, p. 136.

Figure 16. Examples of a Frame filter press (left) and a Chamber filter press (right)

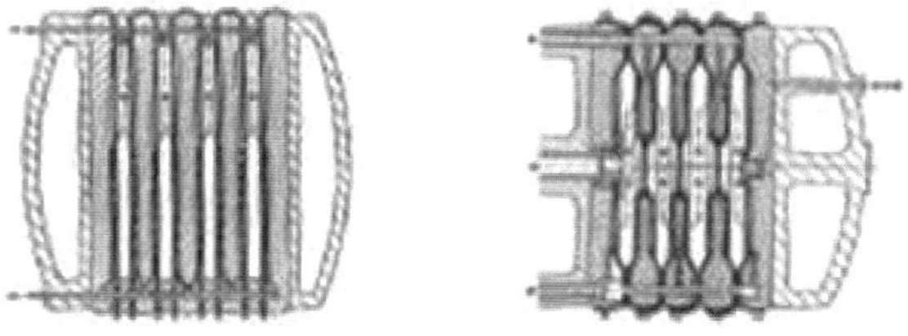

The slurry is pumped into the unit at elevated pressure, and the solids are retained as a cake on the filter cloth attached to the frames. Initially, the filter cloth acts as the filter, but as the cake builds up on the frames, the separated solids perform the bulk of the filtration. The supply pressure is gradually increased to compensate for the increasing resistance of the cake, until the flow of filtrate out of the unit has reached a point where the operation is effectively no longer functioning. At this point, the operation is stopped. In many instances, more filtrate is obtained by blowing air or inert gas through the filter cake and compressing the cake with built-in bladders. The filter cake can then be washed, the filter is disassembled to collect the filter cake, and the frames are cleaned of solids for reuse.

CENTRIFUGAL FILTRATION

These operations are performed in a manner similar to that of a sedimenting centrifuge, save for the fact that the forces at play in filtration, particularly compressible cakes, must be considered. The filtering centrifuge is designed to be operated in batch or semi-batch fashion, given the requirement that the solids must be periodically removed from the unit. Different technologies are available to accomplish this type of operation, the simplest type being the basket batch centrifuge, for which a filter bag is utilized to perform the filtration. However, these units require stopping the centrifuge in order to remove the bag to collect the filter cake. A schematic of the liftout bag discharge is shown in Figure 17.

Figure 17. Liftout bag discharge of a filtering centrifuge[40]

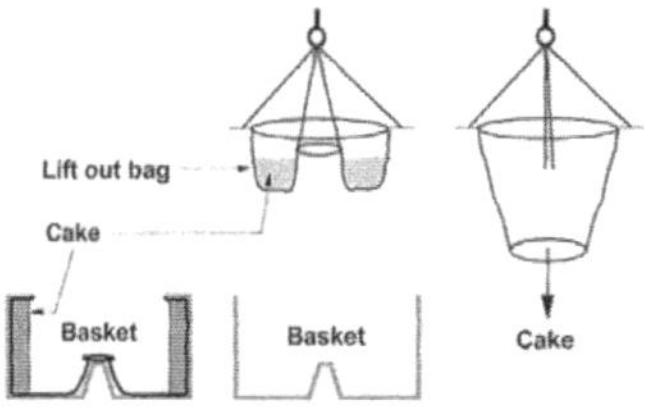

---

[40]Wakeman and Tarleton (Solid/Liquid Separation: Scale-up of Industrial Equipment), p. 329.

There is a considerable operational cost advantage to clearing the filter cakes while maintaining the movement of the centrifuge body. Industry has recognized this need, which has yielded some very innovative designs, such as the peeler centrifuge (solids are scraped from the inside of the bowl once the cake is of sufficient thickness), the inverting bag centrifuge (the moving filter bag is pushed out of the moving basket and discharges the solids when inverted), and the pusher centrifuge (which pushes the solids out the top of the bowl periodically during operation). An example of this type of design is shown in Figure 18.

Figure 18. A filtering pusher centrifuge with solids discharge[41]

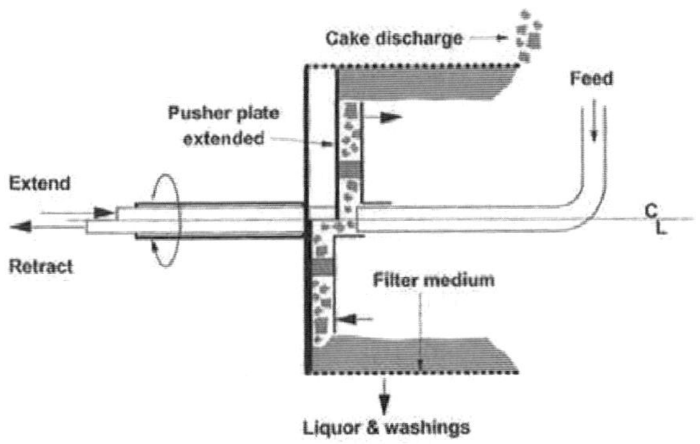

There are many other iterations of this same concept. It should be noted that these designs are superb for their intended use, but they normally come at a very high capital cost.

MICROFILTRATION AND ULTRAFILTRATION

When a solid–liquid separation requires removing extremely small particles, microfiltration or ultrafiltration is a good option. In general, the pore size used to describe these systems is typically related to the size of the material to be separated.[42] The membranes are arranged in such a way as to maximize the filtrate (referred to as the permeate) rate for the smallest membrane surface area. This manner of operation allows for the removal of a portion of the filtrate while using the retained slurry both as feed for the membrane further downstream and also as a way of sweeping solids from the membrane in an active manner. A diagram of cross-flow filtration is shown in Figure 19.

---

[41]Ibid., p. 341.
[42]Ibid., p. 140.

Figure 19. Tangential flow (cross-flow) filtration[43]

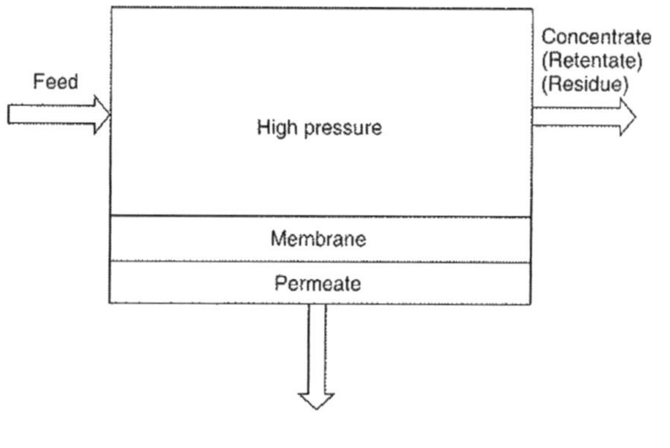

Membranes are arranged into varying geometries based on the needs of the process. A common arrangement is the cassette, which is shown in Figure 20.

Figure 20. A flat sheet membrane assembly used in a cassette[44]

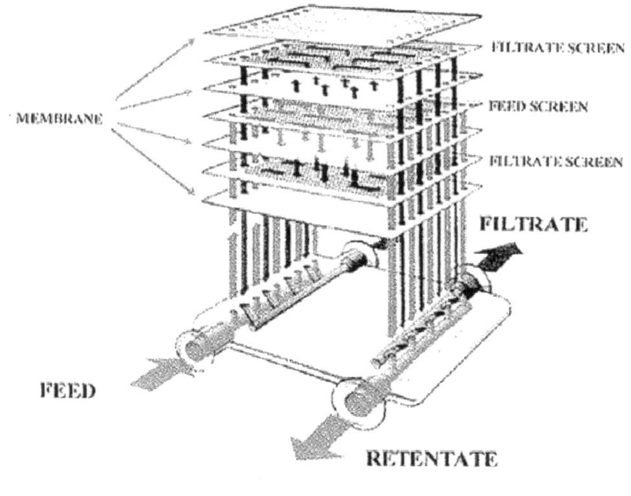

Another common practice is the use of hollow fiber membranes, which contain numerous hollow tubes. The tubular membranes are supported by a pressure vessel and are often run in series. An example of a simple hollow tube module is shown in Figure 20.

---

[43]Perry, pp. 22-38.
[44]Wakeman and Tarleton (Solid/Liquid Separation: Scale-up of Industrial Equipment), p. 158.

Figure 20. Cutaway view of a hollow tube membrane filter module[45]

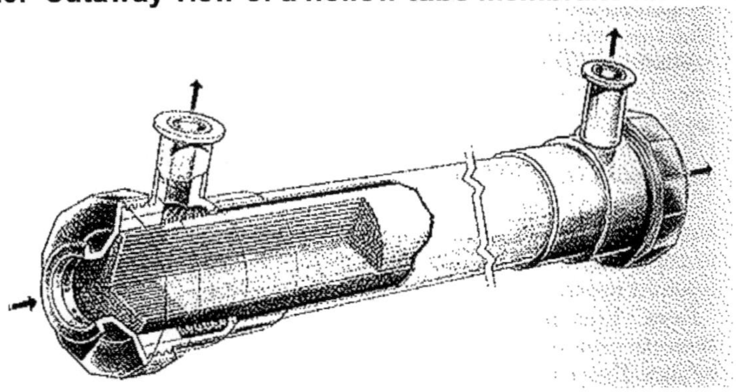

These types of membranes can also be constructed of ceramics, which are typically constructed as monoliths of tubular capillaries selected to serve a particular process separation need. Many of these configurations are plumbed so that a backflow of permeated liquid can be periodically used to push accumulated solids out of the membrane into the feed side to maintain filtration rates. By utilizing proper flows and an effective backflow, membrane filtrations can accommodate a large amount of slurry before a formal cleaning cycle is used to regain the flow through the membranes. It should be noted that although these systems are very effective at removing small amounts of small particles, a heavy solids load on the membranes will cause blinding, which may be irreversible.

CONCLUSIONS

All of the solid-liquid separation technologies in this paper are covered in much greater detail in the references. The theme that one finds with any description of commercial solid-liquid separation equipment is that every technology has its strengths and weaknesses, and often a solid-liquid separation must be evaluated at the pilot scale to verify any equipment selections made from laboratory scale experimentation. Also consistent in the literature is the suggestion that proper separation process design will allow the engineer to take advantage of the benefits of a technology while minimizing those factors that may detract from its perceived value.

The simplest approach to this, in the author's opinions, is to do everything possible to minimize the mass of slurry subject to the solids separation step. Recalling the diagram from Wakeman and Tarleton on p. 4, this requires a development effort in the pretreatment and solids concentration steps in the separation process. Significant benefits to process costs can usually be realized

[45]Perry, pp. 22-40.

by the introduction of low cost/high throughput technologies prior to the use of effective yet expensive solid-liquid separation unit operations.

FURTHER READING

Cheremisinoff, Nicholas, Pocket Handbook for Solid-Liquid Separations: Calculations and Guidelines for Process Engineering, Gulf Publishing Company, 1984.

Evans, Geoffrey M., Atkinson, Bruce W. and Jameson, Graeme J., "The Jameson Cell", Flotation Science and Engineering, Marcel Dekker, Inc., 1995.

Gochin, R.J., "Flotation", Solid-Liquid Separation, 2nd ed., edited by Svarovsky, Butterworths and Co., 1981.

Matis, K.A. and Zouboulis, A.I., Flotation Science and Engineering, Marcel Dekker, Inc., 1995.

Perry, Robert H., and Green, Don W., editors, Perry's Chemical Engineer's Handbook, 7th ed., Sections 18, 19 and 22.

Peters, Max S., Timmerhaus, Klaus D. and West, Ronald E., Plant Design and Economics for Chemical Engineers, 5th ed. McGraw-Hill, 2003.

Pierson, H.G.W., "The Selection of Solid-Liquid Separation Equipment", Solid-Liquid Separation, L. Svarovski, 2nd ed., 1981.

Seidel, D. C. "Laboratory Procedures for Hydrometallurgical-Processing and Waste-Management Programs", United States Department of the Interior Bureau of Mines Information Circular 9431, 1995.

Svarovsky, L., Hydrocyclones, Hold, Rinehart and Winston Ltd., 1984.

Wakeman, R. and Tarleton, S., Solid/Liquid Separation: Principles of Industrial Filtration, Elsevier Advanced Technology, 2005.

Wakeman, R. and Tarleton, S., Solid/Liquid Separation: Scale-up of Industrial Equipment, Elsevier Advanced Technology, 2005.

Warring, R.H., Filters and Filtration Handbook, 1st ed. Gulf Publishing Company, 1981.

AIChE 2010 Annual Meeting, November 7-12, Salt Lake City, UT.

Reactive distillation column for esterification of lauric with methanol: equilibrium vs. nonequilibrium approaches

Nghi Nguyen, Yaşar Demirel*
Department of Chemical and Biomolecular Engineering, University of Nebraska Lincoln, Lincoln, NE 68508

ABSTRACT

A reactive distillation is developed for esterification of lauric acid with methanol using solid catalyst. The reactive distillation column consists of two separation zones and a reaction zone. A super-acid catalyst such as sulfate zirconia is the most suitable candidate for reactive distillation application as it has similar acidic strength to sulfuric acid and highly selective even at high alcohol/acid ratios. However, high content of the water in the reaction zone causes liquid-liquid separation leading to solid catalyst deactivation and inhibition to esterification. In this study, we propose a novel approach that effectively removes the water for the equilibrium and nonequilibrium designs of the esterification of lauric acid. The simulation involves sensitivity analysis of molar reflux ratio and side heater duty on the water concentration in the reactive zone and ester purity in the product stream. Correct placement of side reboiler not only removes water effectively, but also enhances conversion rate by increasing the reaction zone temperature and reduces exergy loss considerably.

Keyword: Reactive distillation column, nonequilibrium distillation, esterification, sulfate zirconia

*Corresponding Author. Tel: 402 472 2745, fax:402 472 6989.
E-mail address: ydemirel2@unl.edu

1. Introduction

Separation by distillation is energy intensive process. Fatty acid esterification by reactive distillation (RD) combines separation and reaction into a single vessel to minimize operation and equipment costs by decreasing waste and overcoming thermodynamic and chemical equilibrium limitations [1-4]. However, usage of RD requires reaction to take place preferably in liquid-phase with a reasonably fast rate, temperature to occur within the suitable separation range, and components with suitable relative volatilities [5].

The well-known equilibrium modeling uses MESH equations to model the equilibrium stages [6]. MESH equations consist of material balances, equilibrium relationships, summation equations, and heat or enthalpy balances. Efficiency approximations are used to compensate for deviation from equilibrium. Equilibrium modeling dominated during last few decades due to their straightforward mathematical modeling. In reality, separation depends on the heat and mass transfer rates between liquid and vapor phases and a more sophisticated nonequilibrium modeling is more suitable to describe the separation process [7-9]. Nonequilibrium modeling uses MERSHQ equations, which composed of material balances, energy balances, heat and mass transfer rate equations, summation equations, hydraulic equations, and equilibrium equations. The results obtained from the equilibrium-based model are used as the initial guess to solve the MERSHQ equations using Newton's method. Nonequilibrium models use Maxwell-Stefan equation to describe the mass transfer between the liquid and vapor phases.

Currently, transesterification/esterification reactions use homogenous catalyst, such as sulfuric acid or sodium hydroxide [10]. The optimum molar ratio of methanol to oil is about 6:1. However, using homogenous catalyst requires additional steps to remove the catalyst [11]. In this study, we have used a heterogeneous catalyst, sulfated zirconia, with the operation temperature ranging between 130 and 200°C. This heterogeneous catalyst exhibits similar catalytic strength to sulfuric acid but it is more environmentally friendly [12-14].

2. Process description and simulation

Figure 1 shows the reactive distillation column using methanol and lauric acid as basic feedstock to produce methyl dodecanoate and water. The column consists of 30 stages with two separation zones at each end and a reaction zone in the middle, and operates with a total condenser and a kettle reboiler. Stream LAURIC consists of 100 kmol/hr of lauric acid, which is heated to 100 °C before entering top of the reaction zone of the column at stage 3. Stream METH consists of 199 kmol/hr of methanol and 1 kmol/hr of water, which is heated up close to methanol's saturation point and fed into the bottom of the reaction zone of the column at stage 29. The ester with a 104 kmol/hr flow rate is the bottom product, while the distillate is the mixture of water and methanol with a flow rate 196 kmol/hr. The activity coefficient model UNIQUAC is used to estimate the phase equilibrium in the column.

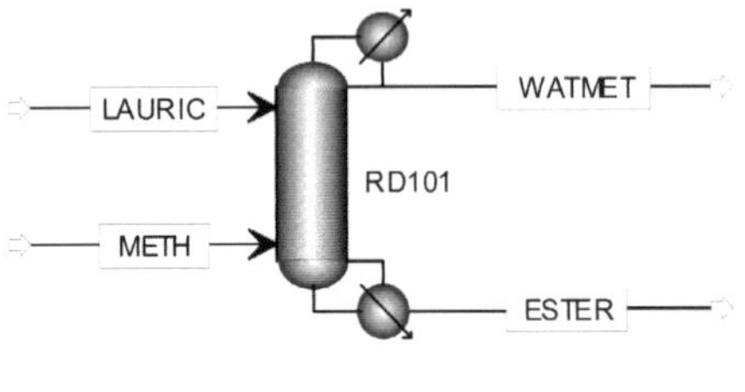

Figure 1: Reactive distillation in the esterification of lauric acid with methanol.

The simulator of Aspen Plus is used to perform the simulation esterification of lauric acid with methanol. Lauric acid comes from coconut oil. The first phase of this work is to reduce the concentration of water in the reaction zone by adjusting the reflux ratio and duty of a side heater in an equilibrium design. The results are then used as initial input values for the nonequilibrium design. A sensitivity analysis has been performed to locate the optimum stage for the side heater. Under identical configurations and conditions, nonequilibrium design has yielded lower conversion due to complex heat and mass transfer rates between the vapor and liquid phases. Therefore, number of stages and pressure needed to be adjusted in order to achieve the same level of conversion. The similar sensitivity analyses for the reflux ratio and side heater duty are also performed for the nonequilibrium design. The rate expression $r = kC_MC_L$ is used for esterification of lauric acid with methanol, where r is the reaction rate, k is the rate coefficient and C_M and C_L are the concentrations of methanol and lauric acid, respectively. The rate coefficient is k = 1.2 10^5exp(-55,000/RT). The units for reaction rate, concentration, and activation energy are kmol/(m^{3}h), kmol/m^3, and kJ/kmol, respectively [3]. The reverse hydrolysis reaction is assumed negligible.

3. Results and discussions

The reactive distillation column operating conditions and configurations are summarized in Table 1. The initial values used for this simulation are shown in equilibrium design 1. In equilibrium design 2, a side reboiler is added in addition to the modifications of feed temperature, molar reflux ratio, and column operating pressure. The modification of pressure is necessary to increase the conversion and to achieve reasonable ester purity at the bottom product stream. However, the main purpose of adjustments in molar reflux ratio and side reboiler is to remove water in the reaction zone, especially at stage 3, since this stage contains highest concentration of water. Additional side reboiler results in less heat duty leading to more economical process due to lower operation temperature. The exergy loss for equilibrium design 2 is 422.3 kW less than equilibrium design 1.

Table 1
Comparison of operating conditions and configurations of equilibrium and nonequilibrium designs for reactive distillation column RD101

Conditions & Configurations	Equilibrium-1	Equilibrium-2	Rate-based-1	Rate-based-2
Number of stages	30	30	30	35
Lauric acid feed stage	3	3	3	3
Methanol feed stage	29	29	29	34
Side reboiler stage	—	3	3	3
Side reboiler duty, kW	0	1200	1200	1200
Lauric feed temperature, °C	100	110	110	110
Methanol feed temperature, °C	90	100	100	100
Molar reflux ratio	1.5	0.7	0.7	0.7
Reaction stages	3-29	3-29	3-29	3-34
Distillate rate, kmol/hr	196	196	196	196
Bottom rate, kmol/hr	104	104	104	104
Bottom ester mass percent	90.23	99.19	95.40	99.17
Liquid holdup, cm^3	0.043	0.043	0.43	0.43
Condenser duty, kW	-5014.33	-3223.82	-3217.17	-3126.79
Condenser temperature, °C	107.92	142.29	141.79	155.39
Condenser pressure, bar	3.5	9.0	9.0	12.5
Reboiler duty, kW	7029.99	4930.54	4968.63	5343.56
Reboiler temperature, °C	224.36	274.34	277.01	297.39
Reboiler pressure, bar	4.0	9.5	9.5	13.0
Total Conversion, kmol/hr	90.19	99.78	95.70	99.75
Exergy loss, kW	1385.5	963.2		

The results of equilibrium design 2 are used as initial input values for rate-based design 1. The heat transfer coefficient, mass transfer coefficient, and interfacial area are estimated using Chen and Chuang, Chilton and Colburn, and Zuiderweg correlation methods, respectively [6,7,9]. Film non-ideality correction is included in the rate-based calculations. Under identical conditions and configurations, rate-based 1 design yields lower conversion as indicated in Table 1. Therefore, in rate-based design 2, the number of stages and pressure are adjusted in order to achieve the desirable conversion as seen in Table 1.

Figure 2 shows the effects of the reflux ratio on the ester mass fraction in the bottom product stream and water mass fraction at stage 3. A decrease in molar reflux ratio decreases the water concentration considerably but will also decrease the ester purity slightly. Figure 2a shows that a logical operating range of the molar reflux ratio for the column is approximately between 0.7 and 1.1. When the molar reflux ratio changed from 1.1 to 0.7, the ester mass fraction in the bottom stream decreased by 1% but the water mass fraction decreased more significantly, by about 4%.

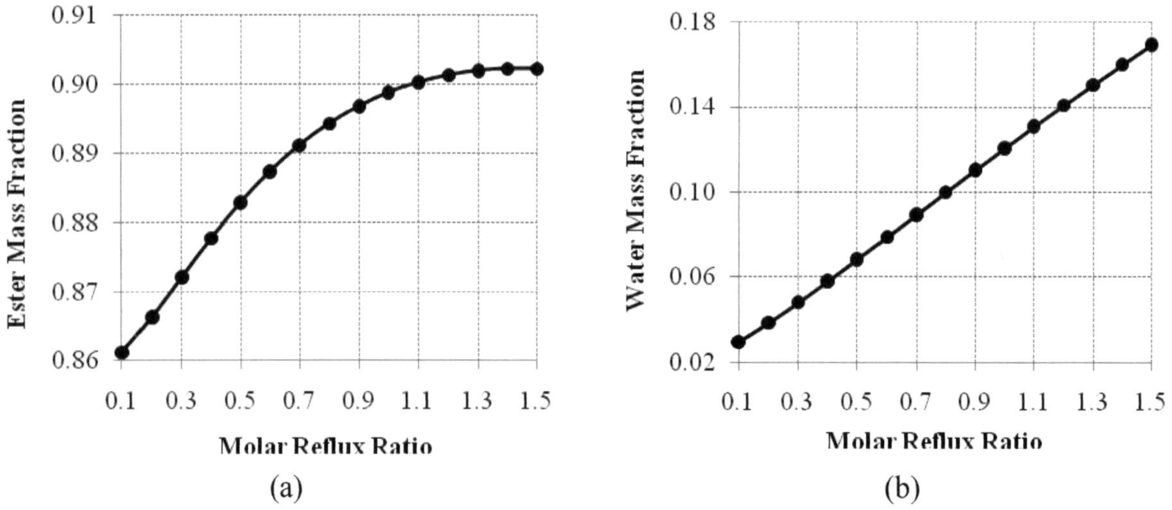

Figure 2. Sensitivity analysis of molar reflux ratio on; (a) ester mass fraction in the bottom product stream; (b) water mass fraction on stage 3.

As the side heater location gets closer to condenser, the water mass fraction of the stage decreases as indicated in Figure 3. The optimal location for the side heater is stage 3. Correct placement of side heater removes water effectively.

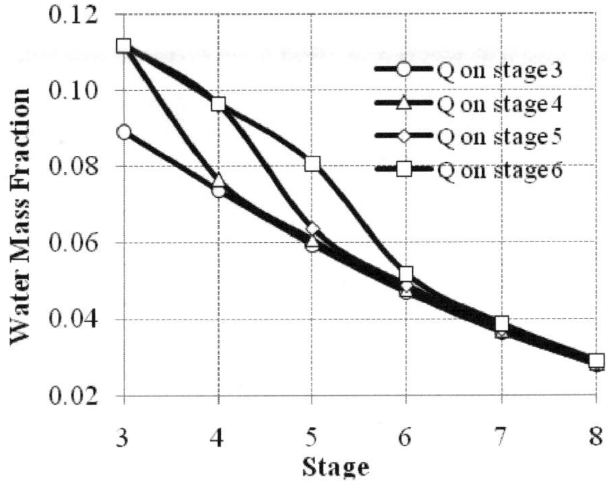

Figure 3. Sensitivity analysis of side heater location on the water mass fraction with a heater duty of 500 kW.

Preliminary analysis indicates that further reduction of water concentration is possible by increasing the side heater duty. Side heater removes water more effectively compared to the reduction of reflux ratio. As shown in figure 4, increasing the side heater duty from 300 kW to 1200 kW results in lower mass fraction of ester in the product stream by approximately 1%, but, the water content on stage 3 decreases more than 4%.

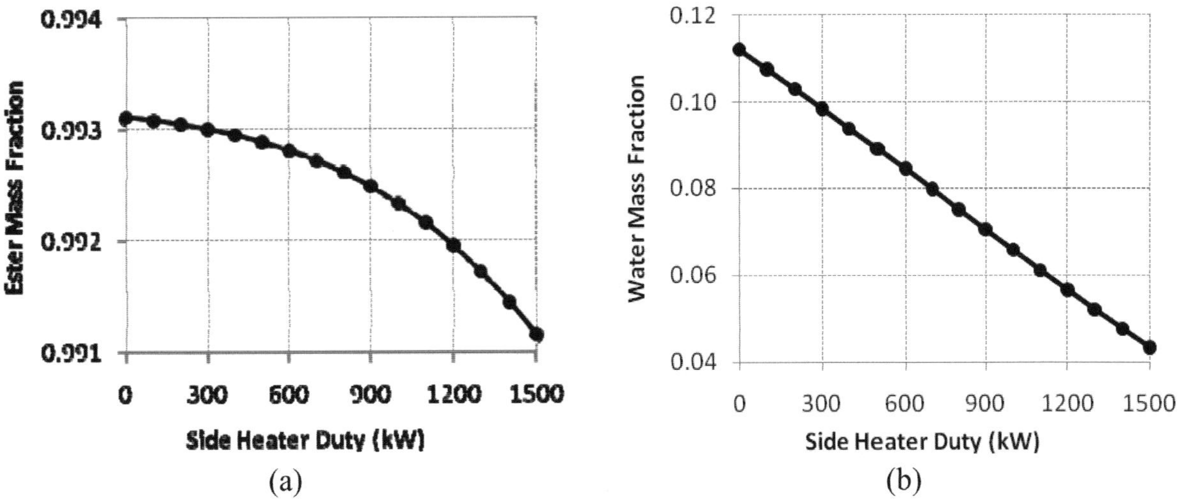

Figure 4. Sensitivity analysis of side heater duty on; (a) ester mass fraction in the bottom production stream; (b) water mass fraction on stage 3.

4. Conclusions

Process simulation of the nonequilibrium design requires more computation effort due to the increased number of iterative calculations. Equilibrium design requires lower operating pressure and number of stages to achieve the required conversion. Sensitivity analysis of reflux ratio and side reboiler duty shows similar trend in both equilibrium and nonequilibrium, designs.

Reflux ratio sensitivity analysis has indicated that reduction in reflux ratio decreases the ester flow slightly in the product stream, however, both the water concentration in the reaction zone and the exergy loss decrease considerably. Correct placement of the side reboiler with the right heat duty has not only removed water effectively but also increased the rate of conversion due to the higher temperature. The reflux ratio modification and side heater addition has shown to effectively remove water in the reaction zone that helps preventing the water inhibiting the reaction.

References

1. Omota, F., Dimian, A. C., Bliek, A. (2003). Fatty acid esterification by reactive distillation. Part 1: equilibrium-based design. Chem. Eng. Sci., 98, 3159-3174.

2. Omota, F., Dimian, A. C., Bliek, A. (2003). Fatty acid esterification by reactive distillation: Part 2 –kinetics-based design for sulphated zirconia catalysts. Chem. Eng. Sci., 58, 3175-3185.

3. Dimian, A. C., Bildea, C. S., Omota, F., Kiss, A. A. (2009). Innovative process for fatty acid esters by dual reactive distillation. Comp. Chem. Eng., 33, 743-750.

4. Kiss, A. A., Dimian, A. C., Rothenberg, G. (2008). Biodiesel by catalytic reactive distillation powered by metal oxides. Energy and Fuels, 22, 598-604.

5. William, L. L., Cheng-Ching, Y. Reactive Distillation Design and Control, Wiley; New York, 2008.

6. Taylor, R., Krishna, R., Kooijman, H. (2003). Real-world modeling of distillation. Chem. Eng. Prog., 99, 28-39.

7. Agarwal, S., Taylor, R. (1994). Distillation column design calculations using a nonequilibrium model. Ind. Eng. Chem. Res., 33, 2631-2636.

8. Higler, A., Krishna, R., Taylor, R. (1999). Nonequilibrium cell model for multicomponent (reactive) separation processes. AIChE J., 45, 2357-2370.

9. Peng, J., Lextrait, S., Edgar, T. F., Eldridge, R. B. (2002). A comparison of steady-state equilibrium and rate-based models for packed reactive distillation columns. Ind. Eng. Chem. Res., 41, 2735-2744.

10. Zhang, Y., Dubé, M. A., McLean, D. D., Kates, M. (2003). Biodiesel production from waste cooking oil: 1. Process design and technological assessment. Bioresource Technol., 89, 1-16.

11. Nguyen, N., Demirel, Y. (2010). Retrofit of distillation columns in biodiesel production plants. Energy, 35, 1625-2632.

12. Lopez, D. E., Goodwin Jr., J. G., Bruce, D. A., Lotero, E. (2005). Transesterification of triacetin with methanol on solid acid and base catalysts. Applied Catalysis A: 295, 97-105.

13. Zabeti, M., Daud, W. M. A. W., Aroua, M. K. (2009). Activity of solid catalysts for biodiesel production: A review. Fuel Processing Technol., 90, 770-777.

14. Kiss, A. A., Dimian, A. C., Rothenberg. G. (2005). Solid acid catalysts for biodiesel production towards sustainable energy. Adv. Synth. Catal., 348, 75-81.

(428e) Zeolite Membranes Supported On Thin Porous Metal Sheet for Water/Ethanol Separation

Jian Zhang, Pacific Northwest National Lab, Richalnd, WA and Wei Liu, Energy and Environmental Directorate, Pacific Northwest National Lab, Richland, WA

This paper investigates the feasibility of hydrophilic and hydrophobic zeolite membranes growth on a unique porous metal sheet of 50um thickness in order to develop low-cost inorganic membrane module of high surface area packing density such as >1000m2/m3. The porous Ni alloy sheet of micrometer or sub-micrometer pore size was prepared by a proprietary process recently developed at PNNL at a controllable thickness over a range of 20 to 1000μm. The hydrophilic and hydrophobic zeolites are NaA zeolite and silicalite zeolite respectively. The membranes were deposited on the porous metal substrate by secondary hydrothermal growth. Effects of seed coating, precursor growth solution, growth conditions, and post-synthesis treatment on the membrane performance were studied. The membranes were characterized by SEM, XRD and water/ethanol separation at various conditions. The hydrophilic membrane, which is water-selective, exhibited a high water/ethanol selectivity of >15,000 and a permeation flux of ~4 kg/m2•h at 75°C with a feed of 10wt% water in ethanol. The membrane also exhibited good stability in 70-h continuous testing at 75°C and 90 °C. The hydrophobic membrane, which is ethanol-selective, exhibited an ethanol/water selectivity of 23 and a permeation flux of 1.7 kg/m2•h at 75°C with a feed of 90wt% water in ethanol.

See more of this Session: Separations for Renewable and Sustainable Energy Development I
See more of this Group/Topical: Topical 1: Separation Needs for Energy Independence and Environmental Sustainability

Continuous Pressure and Vacuum Filtration and Washing of Organic Cellulose and Biomass Products

Barry A. Perlmutter, President & Managing Director
BHS-Filtration Inc.
9123-115 Monroe Road
Charlotte, North Carolina 28270
Phone: 704.845.1190
Fax: 704.845.1902
E-mail: barry.perlmutter@bhs-filtration.com

INTRODUCTION

The processing of organic, cellulose-containing materials has been around for many years as engineers have developed clever chemistry or cultivated natural conversion processes to make products with characteristics, which were not naturally available. Age-old examples are paper, linen or in the field of "conversion" to alcohol and methane.

Currently, due to high oil prices along with the passage of the Renewable Fuels Standard (RFS) included as part of H.R.6, the Energy Independence & Security Act signed in December, 2007 calling for significant increase in ethanol usage nationwide from both grain based and cellulosic based sources, there is tremendous interest in developing efficient, cost-effective manufacturing processes to produce energy from cellulosic feed stocks.

Further, even if the main focus is not on energy production, well-established processes are being changed so that what used to be by-products are now becoming, if not primary products, at least valuable additional products. In addition, the combined effects of an excess of waste and a diminishing supply of basic raw materials focuses the attention of research workers on treating "waste" as a potential raw material for further processing, thus opening a whole new field of bio-processing, bio-energy and specialty chemicals.

This article will use actual process data to highlight the choice of pressure or vacuum filtration and cake washing techniques for these newly developed bioprocesses.

BIOMASS

The word biomass is perhaps misleading since it has different meanings in different industries. Any suitable plant or the products of plants which contain cellulose can be labelled "biomass." Plants include, but are not limited to, trees, bushes, agricultural plants, weeds, vines, straw, flowers, kelp, algae and mixtures thereof. Commercial and agricultural waste products are also included such as stalks, paper, cotton, bagasses, corn cobs, corn stovers, starch, sugar cane, etc. Plants that have been partially decomposed, such as humus, peat, certain soft brown coal etc., may also fall into this category.

Nevertheless, in this article, biomass is defined as a mass of organic solids saturated by, or suspended in, a liquid, which may or may not be organic, and of which it is required that the liquid phase is separated, or replaced by other liquids to enable further processing. It is the filtration / separation in this process stage which is the focus of the discussion. In this article biomass does not mean digested sludge from sewage works.

Regardless of the type of biomass, most bioprocesses rely on reactions of solids and liquids, which later require to be separated. Because no liquid / solid separation process is 100% efficient, it is inevitable that some of the liquid phase will stay embedded in the solid phase and therefore require either mass transfer or evaporation.

The physical form in which the solid content can present itself is almost as varying as there are processes. It may be in the form of finely ground organic matter, it may be granules, fibres, chips or strips. It may be that it has been chemically digested and is amorphous or it may have the original structure. The liquid phase may be water-thin or very viscous. The temperature of the mixture may vary as well as the flow characteristics of the mixture itself may be free-flowing or barely pumpable. Finally, the percent solids and quantity may be large or moderate.

They have, however, one thing in common in that the solids do not like to give up their liquid content very easily and recovering or removal of the process liquors usually involves washing or leaching.

SOLID / LIQUID SEPARATION SYSTEMS

Filtration, pressure or vacuum, is the art of finding or creating a septum or substrate, which allows the liquid to pass through while retaining the solids. The driving force may be gravity, vacuum, pressure or compressive forces whereas a degree of evaporation is often possible using direct heat, indirect heat or electro-magnetic radiation.

The ultimate solid phase is always drier than that from gravitational systems and a great deal of processing can often take place in the actual filtration stage, especially with regard to the conversion or exchange of the residual cake moisture.
The driving force, causing the liquid to exit while leaving the solids behind, can be divided into four groups:

1. Gravitational, which is really little more than draining, but can be very useful to reduce large quantities to more manageable proportions. They can be batch operated or, more usual, continuous. Only a very limited amount of washing of the solids is possible.
2. Vacuum, which is in many ways mechanically the simplest driving force. Vacuum filters can be batch operated but are normally continuous. In general, the solid (cake) thickness can be controlled within close limits; there are few limits to materials of construction and some of vacuum filters offer the best solids washing possibilities. Given that the cake thickness is controllable, residual cake moistures can be obtained which are often equal to pressure systems.

3. Pressure, with or without compression, which obviously involves mechanical constraints. All pressure filters are batch-operated units with one exception, the exception being the BHS Rotary Pressure filter. Cake washing can be excellent and the final cakes are usually as dry as can be expected without heat input.
4. Centrifugal, which can sometimes offer a compromise between vacuum and pressure filtration. Centrifugal filters can be continuous in operation or they can operate in an automated continuous batch-mode. The nature and behavior of the solids play a great part in the success or failure of centrifugal filters. Cake washing can be good depending on the type and the behaviour of the solids.

FILTRATION DRIVING FORCE

Basic Principles

To some extent one would expect that a high pressure driving force would be beneficial for liquid removal, but the effect is limited because organic solids are compressible. In some cases, then, vacuum filtration would be more efficient. As will be discussed later, some cakes can be handled by pressure or vacuum. This is one of the difficulties of biomass filtration.

With a compressible cake, the act of forcing liquid through the already collected solids (the cake) result in these solids being pressed together forming a dense matt. This dense matt, which requires a greater force to penetrate, thus produces a denser matt, which requires even higher forces which produce an even denser matt and so on. This problem multiplies exponentially with the thickness of the cake layer.

Figure 1 illustrates this phenomenon. Normally one can expect a "best overall efficiency" if the cake thickness is kept well below its theoretical maximum and the driving force is not the maximum available. This is true in general, but especially so in those cases where the solids have to be washed, extracted or impregnated subsequently, which is almost always the case with biomass materials.

CAKE WASHING TECHNIQUES

Cake Washing / Extraction / Impregnation

Most biomass processes require cake washing to extract and recover the valuable liquid product. Alternatively, impregnation washing with a liquid, which contains additives in order to, impregnate the solids with these additives, may also be required.

Basic principles

The purpose of cake washing is to remove the residual cake liquor, either because it is valuable or because its presence is undesirable in the cake or both. Depending on the complexity of the cake structure, the possibilities include displacement or surface washing and leach washing.

There may be a third reason for washing, i.e. the wish to impregnate the cake solids with another chemical. The mechanics of this "Impregnation Wash" can be identical to those of the previous three.

A filter cake is made up of particles, which are closely packed together but never perfectly packed together such that there are spaces between them. The spaces form pockets, which are filled with liquor. Depending on how the solids are packed or stacked, a washing liquid is very effective or barely effective.

For stacked particles or puzzle type, the wash fluid will simply run off the particles and only wash off the outer surface but never reach the inter-particle fluid; whereas for more permeable cakes, the wash fluid would have reasonable access to the interstitial spaces, as well as being able to wash off the surface liquor.

However, the actual particle may contain liquor within its own structure, either because it is porous or because it has a cellular structure, as is the case with all undigested organic materials. In this case, the previous processes have hopefully, as much as possible, ruptured the cell walls so that the cellular liquor is mostly accessible. In some cases, however, there may remain a significant amount of cellular liquid, which can only be removed either by heat or osmosis.

Displacement wash

In a displacement wash, the goal is to push the "free" liquor from between the particles out of the cake, using another liquor, which acts like a piston on the former. For this to have a maximum efficiency the two liquids should ideally not be miscible. The success of a displacement wash is normally more a function of the right type of filter, than of the nature of the wash liquors.

Surface wash

However, even under ideal conditions the displacement wash would still leave the particles with a film of diluted process liquor, which would require a surface washing.

Washing times

The former two types of washing are barely time dependent, in so far that their time cycle is the time required to pass "X" volume of wash liquor through the cake at "Y" pressure, which in turn depends on the point on the filtration curve where the washing takes place. From the filtration curve in Figure 1, two (2) volume units would pass at the "optimum" point in one time unit, whereas at point time unit = 6, it would be ten time units. Increasing the pressure normally does not help because of the greater compaction of the cake resulting from the higher filtration pressure.

Leaching Wash with Counter-Current or Reflux Options

As the name implies, a leaching wash attempts to leach liquors held inside the particles into a leaner liquor. In the case of cellular materials this is really closer to an osmotic or an absorption process than a "washing" process. Because the absorption rate is usually quite small, large quantities of liquor have to be passed through the cake, which would result in large quantities of very dilute wash extract. To compensate for this a counter current washing train may well offer the answer, as shown in Figure 2.

Like any counter current system, the reactant will flow in the opposite direction of the main body so that wash liquor meets increasingly "richer" cake from which it absorbs some of the original liquor and thus becomes increasingly "richer" itself. Very considerable savings in

wash volumes can be made this way, although it must be remembered that each step is part of the filtration process and thus requires time and therefore filter area.
It must further be understood that the efficiency is largely dependent on the ease with which the original liquor leaches into the wash liquor and in general the leaching is more effective into lean wash liquor than into liquor, which already carries some of the original liquor.

Alternatively, reflux washing is also possible. In this case, the washing filtrate is recirculated back over the same cake and thereby increasing the efficiency of the wash.

Reslurry Wash

There are, unfortunately, cases where the transfer rate is so slow that it would be senseless to try to carry out a leaching wash in a filter, since it would take up expensive process equipment for inordinate times. In this case the most economical solution is to discharge the filter cake after initial washing and after having reached a point where washing is no longer a function of filtration technology but simply a function of contact time. The cake then gets discharged, mixed with wash liquor to form a slurry and enters into a suitable soaking / leaching system.

DRYING DISCUSSION

Unfortunately, the filtration industry has a habit of talking about "dry" filter cakes. The very non-specific criterion is usually "can the cake be handled by mechanical means, like conveyors, scrapers etc. without losing its shape or without "free" liquid. Therefore even "dry" filter cakes are never really dry, but always contain some or a lot of moisture.

Nevertheless many processes ultimately require bone-dry solids. In general it must be accepted that a filter is not a dryer and that its basic concept is almost the exact opposite of a dryer. It is therefore suggested that if at all possible the drying phase should be regarded as separate from the filtration. However, there are clearly cases where the possibility of being able to process the incoming suspension through all stages of dewatering, washing, extraction, impregnation and also drying would offer tremendous advantages. This may be reasons of hygiene, safety, health hazards, product degradation etc. For such cases a small number of filters can offer an incorporated drying step.

TYPICAL LABORATORY TESTING TO DETERMINE THE OPTIMUM VACUUM OR PRESSURE FILTRATION TECHNOLOGY FOR AN ACTUAL BIOMASS MATERIAL

Overview of Bench Top Testing

The BHS bench top testing is conducted using the BHS Pocket Leaf Filter, as shown in Figure 3. The testing will analyze cake depths, operating pressures, filter media, washing and drying efficiencies and qualitative cake discharge. The data collection sheets are shown in Figure 4.

1. Test purpose

Process description:

Biomass material shall be used as feed material for the production of a bioenergy product. The filter cake from the acidic slurry is reslurried with caustic soda solution and after heat treatment for a certain period of time; the slurry shall be filtered, either by vacuum or pressure filtration. The filter cake shall be washed with the wash filtrate being recirculated two times (three step counter current washing). The wash ratio shall be determined during the testing. Therefore the samples of filtrates and filter cake are collected and analysed. Specific limits for the content of impurities or product in the filter cake shall be determined separately and are not part of this testing.

Requirements:

Production capacity:	20 tons of dry solids / 24 hours
Slurry composition:	2 – 12 % solids per weight
Washing:	Three step, counter current, to be confirmed
Washing media:	DI water
Residual moisture:	As low as technically possible on the filtration selection

Objectives:

- Choice of a suitable filter cloth for this process
- Determine the vacuum or pressure filtration properties of the biomass
- Test several wash ratios for the washing of the filter cake
- Determine the moisture content of the filter cake

2. Test Equipment

The test device is a BHS pocket leaf filter with a filter area of 20 cm^2 and a vacuum and pressure connection. The testing was done at the bioenergy pilot plant. The slurry was prepared for each test using solid material that had been taken from the pilot process. All testing was done at ambient temperatures of 25°C.

3. Test Data and Test Results

While the test data is confidential, the lab testing of the biomass material for the filtration step has shown that the slurry can be easily processed on a continuous working BHS filter. Both vacuum and pressure tests were conducted and gave good results regarding filter cloth selection, filtrate clarity, filtration times, washing parameters and cake discharge.

4. Scale Up for Vacuum and Pressure Filtration

A. Vacuum filtration

The required production capacity is set to 835 kg/h of dry solids. This results in an amount of slurry of approximately 14 tons/ hours at 6% solid content. For the calculation, the wash ratio 1.5:1 is considered optimum based upon the testing.

Filtration Flux rate x:

$$x = \frac{m}{A_F * \sqrt{t_F}} = \frac{6}{20 \cdot \sqrt{10s}} \frac{g}{cm^2} = 0{,}95 \frac{kg}{m^2 s^{0,5}}$$

Based upon the above flux rate, the ratios of filtration areas, washing areas and drying areas are calculated. The result is a BHS continuous–indexing vacuum belt filter with a total area of 16 m² and eight (8) zones. The zone design is shown below.

Process Step	**Number of zones**
Filtration	1
Washing 1	1
Washing 2	1
Washing 3	1
Drying	3
Spare	1

B. Pressure Filtration

Similar to the scale-up for the vacuum filtration case, the required production capacity is set to 835 kg/ h dry solids. This results in an amount of slurry of approximately 14 tons/ hours at 6% solid content. For the calculation, the wash ratio 1.5:1 is considered optimum based upon the testing.

Specific total filter performance flux rate:

$$Q = 90 \frac{1}{h} \cdot \frac{6g}{20cm^2} = 270 \frac{kg}{m^2 h}$$

Required filter area:

$$A = \frac{835 \frac{kg}{h}}{267 \frac{kg}{m^2 h}} = 3{,}1\ m^2$$

The result is a BHS rotary pressure filter a total filter area of 3.2 m², with an arrangement shown below.

Process Step	**Degrees of Rotation**
Filtration	57
Washing 1	57
Washing 2	57
Washing 3	57
Blow Drying	57
Cake Discharge and Cloth Cleaning	75

5. Selection and Discussion of Vacuum and Pressure Filtration

Both of the technologies, the BHS vacuum belt filter as well as the BHS rotary pressure filter can be selected for the processing of the biomass material. Following this section, there is a brief discussion of each of the technologies.

The design of the rotary pressure filter assures that the cake is always saturated in between filtration and the different washing steps. There is no risk of air getting inside the cake and maximum washing efficiency is achieved. Only during the blow-drying step is the moisture in the cake mechanically removed.

On the belt filter, the zone arrangement and positioning of the washing devices on the filter can be used to assure saturated cake during the washing procedure. As another option it is also possible to work with sucking of the liquid in between the washing steps in case that this increases the washing efficiency. The belt filter offers the flexibility to easily adjust the zone arrangement by moving the washing devices.

Finally, due to the higher dewatering pressure difference, the residual moisture of the filter cake is approximately 3 % lower with pressure filtration than with vacuum filtration.

The choice of the preferred technology in this case might also consider the required surrounding components. In case of the belt filter these are mainly the vacuum pump including pressure side separator for the operation liquid, the filtrate separators (three pieces) and the filtrate pumps for the transportation of the wash liquids from the filtrate separators back onto the filter. For the rotary pressure filter, a feed pump for the slurry will be required that is able to provide the filtration pressure, filtrate pumps for the circulation of the wash liquids and a pressure gas system (e.g. compressor, pressure control etc.) to allow for the dewatering of the filter cake.

In summary, the decision resides with the other process parameters of the operation including washing efficiency, downstream equipment as well as the economics of the installation. The client continues the evaluations of both options.

FILTRATION TECHNOLOGIES

High-Solids Slurries: Continuous Vacuum Operation

High-solids slurries can be defined as up to 50 – 55% solids in the slurry feed. An example of a continuous, thin-cake technology is the BHS Continuous – Indexing Vacuum Belt Filter. This technology consists of fixed vacuum trays, continuously feeding slurry system and indexing or step-wise movement of the filter media. The filter media is indexed by pneumatic cylinders located on the exterior of the unit. The pneumatic operation and fixed trays eliminates a motor and variable speed drive, there are no rails/rollers, no rubber carrier belt and flexible pressure-vacuum rated hoses are not required. For cloth movement, the vacuum is broken by butterfly valves and the cloth moves (indexes), by pneumatic cylinders, in the space above the vacuum trays. The belt moves along but never touches the trays, which allows for long filter cloth life.

For the process operation, due to the stepwise operation of the belt, washing and drying efficiencies are maximized, as the belt is stopped and the mechanism of "plug-flow" for gases and liquids is in effect. Drying can be by vacuum, steaming, hot gas and mechanical pressing. Pressing is effective with biomass materials due to thier compressibility. Pressing can be combined with blowing for vacuum-convection drying.
Finally, because the trays are fixed, the mother liquid and the wash filtrates can be recovered individually and recirculated and/or recovered and reused which allows for a more efficient overall operation as previoulsy described in the "Cake Washing Techniques" section.

High-Solids Slurries: Continuous Pressure Operation

In some cases, for high solids applications, the slurries can be better handled using pressure filtration rather than vacuum filtration. The BHS Rotary Pressure Filter technology provides for continuous production in a single unit.

Filtration is conducted via pressure of up to 90 psig. Positive displacement washing, reflux or counter-current washing follows filtration. Of course, multiple washing steps as well as solvent exchanges, steaming and extraction can also be accomplished. Finally, the cake is dried by blowing hot or ambient-temperature gas through the cake.

The Filter has a uniquely designed discharge system, which provides for atmospheric discharge from pressure filtration. After automatic cake discharge, the filter cloth is washed; the clean filter cloth the re-enters the feeding / filtration zone thereby continuing the process. All solvent and gas streams can be recovered separately and reused in the process to minimize their consumption.

SUMMARY

Cellulose and biomass filtration, washing and drying is very unpredictable and even minor differences in feedstock, washing or cake moisture requirements can require a different process solution. As previously described, both vacuum and pressure operations provided basically equivalvent process results. The process decision will then be the combined overall efficiency, total installed capital cost, operating cost, space requirements, ancillary equipment and above all the reliability and cost-effectiveness which will result in the optimum filter selection.

Ultimately, there is no substitute for accurate and professional test work under realistic conditions. This almost certainly means that tests have to be done by the technical staff of the filter manufacturer, preferably one who manufactures both vacuum and pressure filtration technologies, either in their laboratory or at the client's site, since they will have the right test equipment, the expertise of doing the test and the expertise to interpret the results. For this, close cooperation between the plant operations and engineering staff and the filtration vendor is necessary along with detailed and professional laboratory and pilot plant tests.

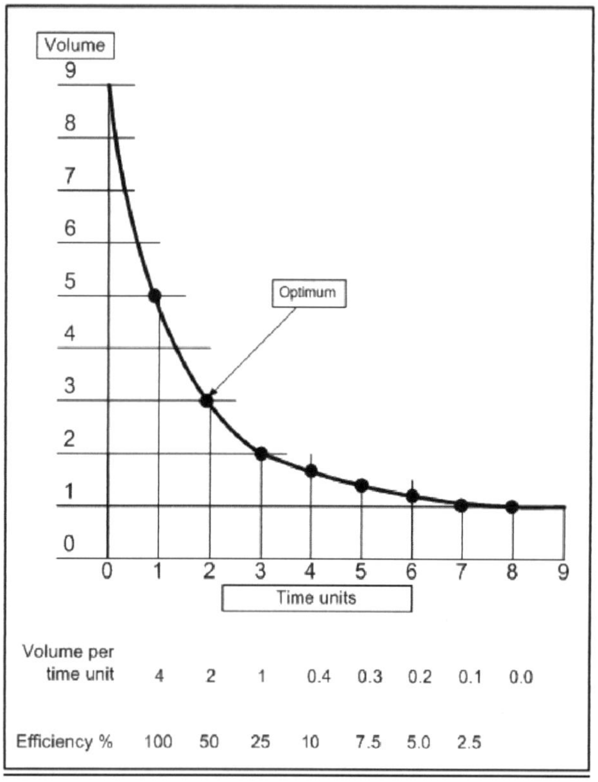

Figure 1: General filtration curve showing as filtration continues, the cake grows thicker and denser, while the efficiency decreases

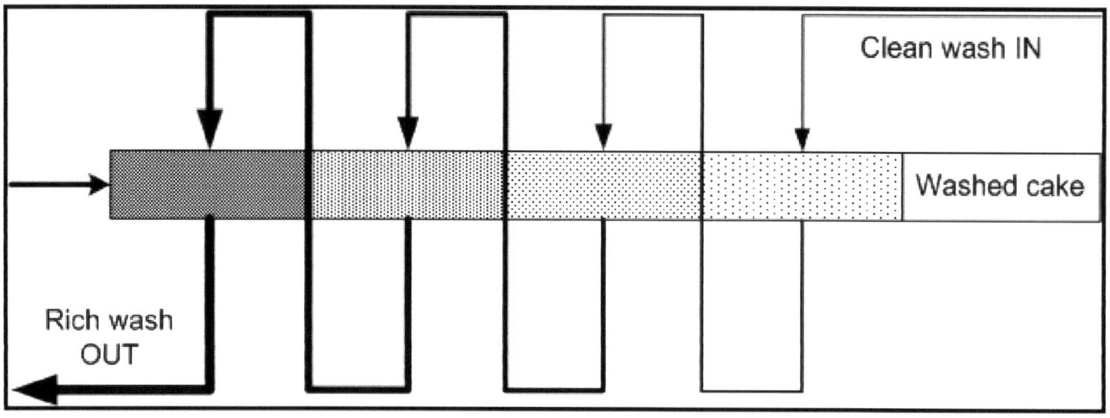

Figure 2: Typical counter current washing train

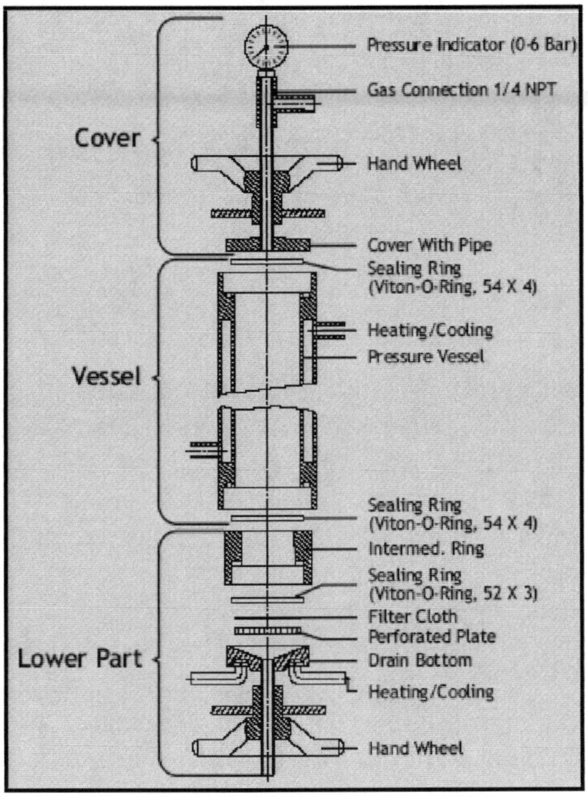

Figure 3: BHS Pocket Leaf Filter

Customer:		Test Number:
Date :		Run #
	Filter Media	
	Suspension	
Filling	Volume of Slurry	
	Density of Slurry	
	% Solids in Feed	
	Temperature	
Filtration	Pressure/Vacuum	
	Volume of Filtrate	
	Time for Filtration	
	% Solids in Filtrate	
Wash 1	Wash Material	
	Pressure/Vacuum	
	Volume of Filtrate	
	Time for Filtration	
Wash 2	Wash Material	
	Pressure/Vacuum	
	Volume of Filtrate	
	Time for Filtration	
Wash 3	Wash Material	
	Pressure/Vacuum	
	Volume of Filtrate	
	Time for Filtration	
Drying	Pressure/Vacuum	
	Temperature	
	Flow Rate	
	Time for Drying	
	Pressing Pressure	
Cake	Weight	
	Thickness	
	% Residual Moisture	
	Dry Cake Weight	
	Cake Discharge OK?	

Figure 4: Data Collection Sheet for BHS Pocket Leaf Filter

Solvent Extraction of Lipids from Microalgae by Application of Pulsed Electric Field (PEF): Is it Feasible?

Mauricio D. Antezana, The University of Kansas, Lawrence, Ks, USA
Susan M. Stagg-Williams, The University of Kansas, Lawrence, Ks, USA
Belinda S. M. Sturm, The University of Kansas, Lawrence, Ks, USA

Introduction

The scientific community has recognized the great potential of microalgae as an alternative source of fuels. The most important characteristics that make microalgae such an attractive alternative for petroleum fuels are: (i) microalgae requires carbon dioxide to perform their basic metabolism, therefore it is considered CO_2 neutral (ii) when compared to other bio-sources for fuels, microalgae's biomass growth rate is significantly higher, and finally (iii) microalgae does not compete against human food supply or water usage like other bio-sources do.

The development of fuels from microalgae is still on its early stages to become a sound and competitive technology. The production of fuels from algae is divided in different fronts: (a) algae growth, (b) harvesting of algae, (c) lipid extraction, (d) and catalytic conversion. Important technological breakthroughs are still needed in all the different stages. This article investigates the usage of an alternative, "greener" solvent (ethyl acetate) instead of the traditional Chloroform during the lipid extraction step. This paper also describes our attempt to utilize PEF (also know as electroporation) in order to intensify this extractions. The algal strain utilized during this research is Ankistrodesmus.

Background of PEF and initial idea on how to apply it to lipid extraction.

Literature search indicates that the efficiency of solvent extraction from algal lipids can be highly dependent to the utilization of pretreatment that rupture the cell membrane[1, 2]. These types of pretreatment vary from mechanical, enzymatic, to utilization of sound waves to enhance the lipid extraction. The novelty of our research is the application of PEF as a means of disrupting the cell membrane to increase lipid extraction.

PEF is a technique in which cell suspensions are exposed to high electric fields for brief amount of time. When exposed to PEF, the treated cells experience an increase in membrane permeability, which provides greater access into (or out of) the cell cytoplasm. The mechanisms of microbial or cell electroporation are not completely elucidated, but it has been postulated that the lipid bilayer membranes form "pores" which causes this permeabilization increase[3].

To date the most frequent application of electroporation has been in the biology, biotechnology, and medicine fields. Some of the applications of this technology involve the introduction of fluorescent probes to microorganism, loading cells with molecules for drug delivery, and transfection of cells. All the above mentioned applications are of small scale and intend to gain access into the microorganism. A different application of PEF is the inactivation of microbial organism in products like milk[4], apple juice[5], and other liquid products for human consumption. For this type of application the fundamental intent is to irreversibly permeabilize cell membranes to such an extent that microorganism death is produced. The application of PEF with this purpose has been proven to be of industrial scale, and it raises the questions if PEF can be applied to the intensification of algal lipid extraction.

Bligh and Dyer Method & Ethyl Acetate

The Bligh and Dyer Method[6] or slight variations from it, are the most common extraction of total lipids from microalgae for lab scale. In this method chloroform (in conjunction with methanol) is utilized as the main solvent, and it is considered a point of reference for lipid extraction efficiency. However, chloroform presents obvious limitation at the time of scaling up any process, since it represents a health hazard and it is not environmentally friendly. For this research, ethyl acetate was investigated as a potential replacement for chloroform.

Figure 1. presents a typical comparison of the extraction efficiency of ethyl acetate compared with chloroform. Ethyl acetate extraction of algal lipids was found to have 80-92% efficiency when compared to chloroform. This efficiency reduction clearly presents an intensification opportunity.

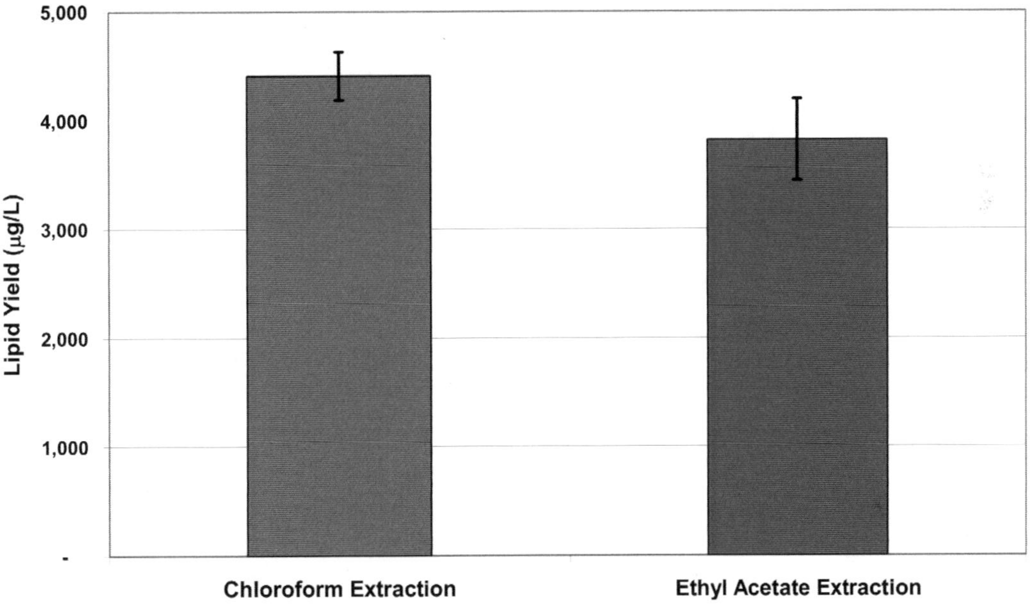

Figure 1. Lipid extraction from microalgae utilizing ethyl acetate are normally 80-92% as efficient as when chloroform is used. The graph corresponds to the average ± 95% confidence interval. The lipid yields are expressed as microgram of lipid per liter of algae solution obtained from gas chromatography runs.

The rate at which algal lipids are extracted in the chloroform system as compared to the ethyl acetate system was also determined. **Figure 2** illustrates that the algal lipid extraction with chloroform solvent is extremely fast, having no time dependency. On the other hand, ethyl acetate solvent has approximately a 60-90 min. time dependency before reaching its maximum extraction. This extraction rate reduction also represents an intensification opportunity.

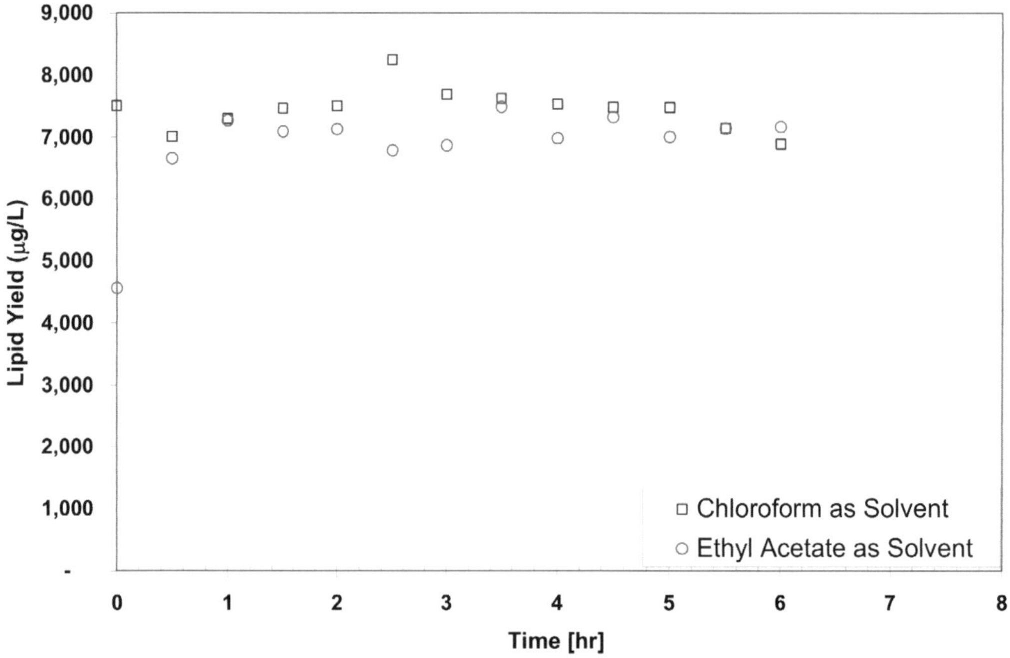

Figure 2. Microalgae lipid extraction with chloroform presents an almost immediate extraction. No significant time dependency can be observed on this system. On the other hand, extraction of lipids with ethyl acetate presents approximately 60-90min time dependency.

Application of PEF technique to Algal cells & Fluorescent Microscopy

The increase of membrane permeability associated with PEF application could translate into a better contact between solvents and the algae cytoplasm. In other words, this permeability increase could correspond to a mass transfer enhancement for lipid extraction, and the type of intensification desired. In order to test this hypothesis and 'In-House' unit, capable of delivering the required pulsed electric field was developed.

In order to confirm that the application of PEF has gained access through the cell membrane, fluorescent molecular stains were used to determine cell viability: Calcein AM and Propidium Iodide (PI). The non-fluorescent Calcein AM is a membrane-permeable dye that loads into healthy cells and at this point is converted into green-fluorescent Calcein. The leakage of this dye is an indicator of membrane permeabilization. On the other hand, PI is a red-fluorescent dye which is not permeant to live cells and only bounds to RNA and DNA of compromised cells.

Figure 3 presents a negative control sample of algae loaded with Calcein-AM and PI, to which no PEF has been applied. This figure clearly indicates the presence of healthy algae since the Calcein signal is present and no PI can be observed. On contrast, **Figure 4** illustrates the effect of algae exposed to PEF, where cell membrane has been compromised. This conclusion can be supported by the lack of Calcein signal and the clear presence of PI.

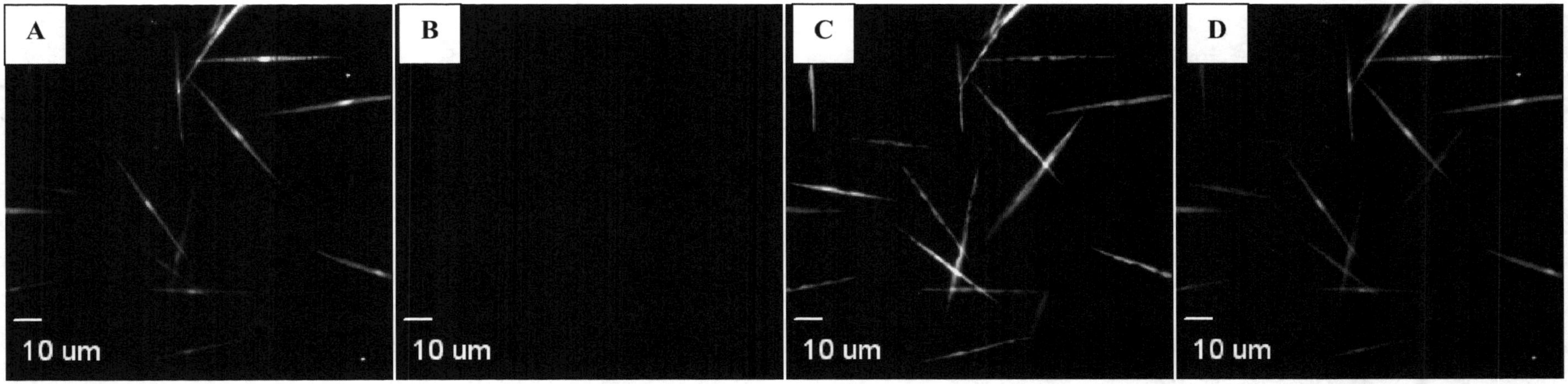

Figure 3. Multi-channel epifluorescence microscopy image of Calcein AM-loaded algal cells with PI present. The different channels are as follows: A) calcein, B) PI, C) algae auto-fluorescence, D) combined channels. This image represents a control negative sample, where the algae has not been exposed to PEF.

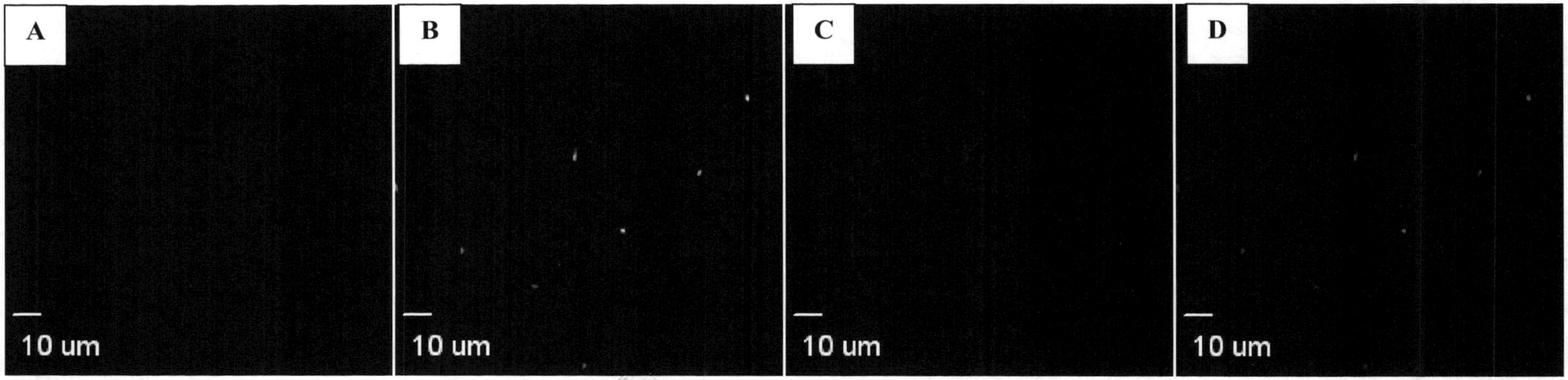

Figure 4. Multi-channel epifluorescence microscopy image of Calcein AM-loaded algal cells with PI present. The different channels are as follows: A) calcein, B) PI, C) algae auto-fluorescence, D) combined channels. This image indicates the permeabilization of the algal cells once exposed to the PEF. Lack of Calcein signal on channel 'A' and presence of PI signal on channel 'B' confirms that our 'In-House' PEF unit successfully works.

The utilization of ethyl acetate as an alternative, 'greener' solvent for algal lipid extraction has shown interesting potential. However, the efficiency of this solvent could be intensified in two areas:

1. Increase of total lipid extraction to achieve chloroform-like levels (or even higher)
2. Improvement of lipid extraction rates, such that the liquid extraction could be immediate, without any time dependency.

At this point it is suggested that the utilization of PEF could translate into an effective technique for cell membrane disruption, which could improve the extraction and mass transfer process. If proven correct, this technique could address the above-mentioned improvement opportunities, intensifying the extraction of lipids for fuel production.

Acknowledgments

We would like to acknowledge the Microscopy and Analytical Imaging Lab for providing support with the imaging acquisition, the Transportation Research Institute for providing resources, and electrical engineer Wes Ellison from Aerospace Engineering for providing support with the development of the PEF unit; all based at The University of Kansas.

References

1. Lee, J.-Y.; Yoo, C.; Jun, S.-Y.; Ahn, C.-Y.; Oh, H.-M., Comparison of several methods for effective lipid extraction from microalgae. *Bioresour. Technol.* **2009,** *101* (Suppl. 1), S75-S77.
2. Ranjan, A.; Patil, C.; Moholkar, V. S., Mechanistic Assessment of Microalgal Lipid Extraction. *Ind. Eng. Chem. Res.* **2010,** *49* (6), 2979-2985.
3. Neumann, E.; Kakorin, S.; Toensing, K., Principles of membrane electroporation and transport of macromolecules. *Methods Mol. Med.* **2000,** *37*, 1-35.
4. Shamsi, K.; Versteeg, C.; Sherkat, F.; Wan, J., Alkaline phosphatase and microbial inactivation by pulsed electric field in bovine milk. *Innovative Food Sci. Emerging Technol.* **2008,** *9* (2), 217-223.
5. Schilling, S.; Schmid, S.; Jager, H.; Ludwig, M.; Dietrich, H.; Toepfl, S.; Knorr, D.; Neidhart, S.; Schieber, A.; Carle, R., Comparative study of pulsed electric field and thermal processing of apple juice with particular consideration of juice quality and enzyme deactivation. *J Agric Food Chem* **2008,** *56* (12), 4545-54.
6. Bligh, E. G.; Dyer, W. J., A rapid method of total lipid extraction and purification. *Can J Biochem Physiol* **1959,** *37* (8), 911-7.

(493b) Alkaline Hemicellulose Extraction and Solubility-Based Separation From Hardwoods

Ryan J. Stoklosa, Chemical Engineering & Materials Science, Michigan State University, East Lansing, MI and David Hodge, Chemical Engineering and Materials Science, Michigan State University, East Lansing, MI

Woody biomass presents a unique opportunity for the production of liquid biofuels and other chemicals from renewable resources. By utilizing woody biomass, existing supply chains and infrastructure developed for the forest products industry can be employed which both shortens the timeline for commercialization for these technologies and greatly enhances the economic attractiveness of pretreatment technologies that can successfully work on these feedstocks. The challenge is to integrate existing alkaline pulping processes with extraction processes that utilize the hemicellulose and lignin fractions for fermentation or catalyst-derived transportation fuels, solid fuels, and chemicals. An alkaline pre-extraction before pulping or black liquor removal during pulping can permit use of the hemicellulose and lignin fractions for higher value fuels and chemicals. For this work, alkaline extractions will include variations in NaOH loading, time and temperature, and biomass source. The biomass feedstocks will include 3 hardwoods: sugar maple (Acer saccharum) and hybrid poplars (Populus tremuloides x tremula, P. nigra x maximowiczii NM6). Extracted and solubilized hemicelluloses and lignins exhibit different functionalities due to their chemical alteration from the original biopolymer and this work will better characterize the impacts of the extraction conditions on the properties of the solubilized biopolymers and the potential for solubility-based separations to recover hemicellulose and lignin fractions using precipitation by acid and organic solvents. The properties of the total solubilized material and precipitates as a function of extraction condition will be characterized in terms of overall composition, hemicellulose and lignin molecular weight distributions by size exclusion chromatography and a novel assay for xylan degree of polymerization estimation, the degree of substitution of xylan polymers with 4-O-methyl glucuronic acid and acetate, and hemicellulose-lignin aggregation.

See more of this Session: Separations for Renewable and Sustainable Energy Development II
See more of this Group/Topical: Topical 1: Separation Needs for Energy Independence and Environmental Sustainability

Evaluation of Immobilized Boronic Acid for Shifting the Xylose:Xylulose Equilibrium with Immobilized Xylose Isomerase

Bin Li[1], Dawei Yuan[1], Heng Shao[1], Patricia Relue[1] and Sasidhar Varanasi[2], (1)Bioengineering, University of Toledo, Toledo, OH, (2)Chemical & Environmental Engineering, University of Toledo, Toledo, OH

Saccharification of lignocellulosic feedstocks to sugars can result in bimass hydrolysate that is up 30% xylose. Native S. cerevisiae (baker's yeast) is not able to ferment xylose to ethanol; this yeast can ferment xylulose, the ketose isomer of xylose, to ethanol. D-glucose isomerase (D-xylose isomerase, EC 5.3.1.5) is known to isomerize xylose to xylulose. However, the chemical equilibrium for this reaction is unfavorable for xylulose production, with an equilibrium xylulose:xylose ratio of ~3:7 at neutral pH 7 and 50 °C.

Boronic acids (with a general structure of $R\text{-}B(OH)_2$, where R is an alkyl or aryl) have been known for some time to form boronate esters with diols rapidly and reversibly in aqueous solution. The affinity between different boronic acids and diols in aqueous solution is pH dependent, with maximal binding at high pH values (> pH 8) and diol release at low pH (pH< 3). Borax, or sodium tetraborate, was shown to significantly improve the xylose:xylulose equilibrium in sugar solutions and in biomass hydrolysate. However, tetraborate in high concentrations can adversely impact the growth of some microbes used for the fermentation of biomass sugars.

To determine if other boronic acids may be better candidates for shifting the xylose:xylulose equilibrium with less inhibition of microbial physiology, different boronic acids were investigated. The optimum pH for xylose conversion was determined for each and the optimal ratio of sugar to boronic acids was also determined.

Several promising boronic acids were immobilized to a solid support to further improve process economics and to avoid cell growth inhibition from those chemical compounds. Results for the sugar recovery and isomerization with free and immobilized boronic acids are also presented.

(493d) Influence of Membrane Structure and Permeability On Power Production in Pressure Retarded Osmosis Processes

Zeta L.T. Yu-Peralta, Department of Civil and Environmental Engineering, University of California, Los Angeles, Los Angeles, CA and Eric M. V. Hoek, Dept. of Civil and Environmental Engineering, University of California, Los Angeles, Los Angeles, CA

The concept of pressure-retarded osmosis (PRO) was first introduced by Norman in 1974 [1], but later named as PRO [2] and extended by Loeb [2, 3] and other researchers [4-8] . The main obstacle that hinders the development of PRO and limits power production is the lack of appropriate membrane design¬—specifically, a highly permeable and selective barrier layer and a support layer that produces minimal internal concentration polarization (ICP). Internal concentration polarization results from hindered diffusion of feed and/or draw solutes through the porous support layer of a composite membrane. The extent of ICP is theoretically a function of the porous support layer thickness and (macrovoid) porosity and tortuosity. So far, the highest output reported is 3.5 W/m2 of membrane area from a thin film composite membrane tested in the laboratory using simple salt solutions and converting the measured flux into power using a simple thermodynamic model [9]. In order for PRO to be economically viable, a membrane that generates at least 5 W/m2 is required [10]. This goal of the work to be presented is to elucidate the inter-relationships between the five known PRO structure-performance parameters (barrier layer water permeability and salt permeability; support layer porosity, tortuosity, and thickness) and their influence on PRO power production. In the talk, we will map out various combinations of the five structure-performance parameters required to make PRO economical using the commercial CTA membrane properties as the baseline for comparison. Our preliminary results suggest that for a given porous support layer increasing the barrier layer water permeability directly increases power density; however, if salt permeability increase proportionately the gain in power production is completely lost. Hence, PRO membranes must become more permeable to water, while maintaining the current salt permeability. For a given barrier layer, support layer tortuosity appears the most important parameter; as tortuosity approaches unity, power production is minimally affected by changes in support layer porosity and thickness. These results provide great insight into the structural changes required to produce a viable PRO membrane.

References

1. Norman, R.S., Water Salination - Source of Energy. Science, 1974. 186(4161): p. 350-352. 2. Loeb, S., Osmotic Power-Plants. Science, 1975. 189(4203): p. 654-655. 3. Loeb, S., Production of Energy from Concentrated Brines by Pressure-Retarded Osmosis .1. Preliminary Technical and Economic Correlations. Journal of Membrane Science, 1976. 1(1): p. 49-63. 4. McCutcheon, J.R. and M. Elimelech, Influence of concentrative and dilutive internal concentration polarization on flux behavior in

forward osmosis. Journal of Membrane Science, 2006. 284(1-2): p. 237-247. 5. Seppälä, A. and M.J. Lampinen, Thermodynamic optimizing of pressure-retarded osmosis power generation systems. Journal of Membrane Science, 1999. 161(1-2): p. 115-138. 6. Post, J.W., et al., Salinity-gradient power: Evaluation of pressure-retarded osmosis and reverse electrodialysis. Journal of Membrane Science, 2007. 288(1-2): p. 218-230. 7. Xu, Y., et al., Effect of Draw Solution Concentration and Operating Conditions on Forward Osmosis and Pressure Retarded Osmosis Performance in A Spiral Wound Module. Journal of Membrane Science, 2009. In Press, Accepted Manuscript. 8. Achilli, A., T.Y. Cath, and A.E. Childress, Power generation with pressure retarded osmosis: An experimental and theoretical investigation. Journal of Membrane Science, 2009. 343(1-2): p. 42-52. 9. Gerstandt, K., et al., Membrane processes in energy supply for an osmotic power plant. Desalination, 2008. 224(1-3): p. 64-70. 10. Skilhagen, S.E., J.E. Dugstad, and R.J. Aaberg, Osmotic power - power production based on the osmotic pressure difference between waters with varying salt gradients. Desalination, 2008. 220(1-3): p. 476-482.

See more of this Session: Separations for Renewable and Sustainable Energy Development II
See more of this Group/Topical: Topical 1: Separation Needs for Energy Independence and Environmental Sustainability

(493e) Separation and Recovery of Ionic Impurities From Biomass Hydrolysate by Electrodialysis/Electrodeionization

Saurav Datta, Seth W. Snyder, Michael P. Henry, Sabeen F. Ahmad and Yupo J. Lin, Argonne National Laboratory, Lemont, IL

Electrodialysis (ED) is an electricity-driven membrane separation technique capable of removing ionic species from solution. Electrodeionization (EDI) is a modified version of ED that contains ion-exchange (IX) resin beads within it. EDI is superior over ED under certain conditions as the phenomenon of water splitting on the surface of the IX beads enhances the ionic conductivity of the solution. EDI is also more efficient than conventional IX column because of the in-situ regeneration of the IX resin beads under the applied electric field. ED and EDI are used primarily to produce ultra-pure water for the semiconductor and pharmaceutical industries. Researchers at Argonne National Laboratory have improvised the conventional EDI technology (loose IX resin beads) to resin wafer (RW) based EDI (RW-EDI) by developing "IX resin wafers", where IX resin beads are immobilized onto a porous matrix. This novel RW-EDI enables implementation of EDI to wider range of applications than the conventional EDI. Lignocellulosic biomass is one of the major sources of biofuel, which can potentially replace petrochemical based fuel, with a future goal of producing 22 billion gallon of cellulosic ethanol per year. In a typical process, the lignocellulosic biomass is subjected to an acid pretreatment step to rupture the crystalline structure and aid the downstream enzymatic hydrolysis of polysaccharides to sugars. Unfortunately, the pretreatment step generates significant amount of non-ionic (furfural, hydroxymethyl furfural, etc.) and ionic (acidic acid, etc.) impurities along with the residual sulfuric acid that inhibit the downstream processes. The existing sulfuric acid removal technique (limation) is inefficient and requires additional care for handling gypsum. The acetic acid removal techniques are also inadequate in commercial scale. We, at Argonne National Laboratory, are trying to improve the performance of various separation techniques associated with the biomass pretreatment step in order to make the cellulosic ethanol production more efficient. In this context, we have evaluated the application of ED/EDI for the removal of ionic impurities (both sulfuric and acetic acid) from biomass hydrolysate. Both ED and EDI have been used to study the separation of the acids from the synthetic hydrolysate liquor. Different process parameters, such as, flow rate, applied voltage and composition of resin-wafers have been varied to determine the optimum conditions. It has been observed that more than 99 % of the acids could be removed from hydrolysate using ED or EDI. The sugar retention was more than 95 %. Further attempt has been made to recover the sulfuric acid and acetic acid in purified form so that the sulfuric acid could be recycled and the acetic acid could be utilized in other applications.

See more of this Session: Separations for Renewable and Sustainable Energy Development II
See more of this Group/Topical: Topical 1: Separation Needs for Energy Independence and Environmental Sustainability

Results from the Industrial Water Use Efficiency and Desalination Workshop

Bhima Sastri, Ph.D. ,Technology Manager, ITP, DOE, Washington DC
Fred Hansen, Ph.D., Senior Consultant ,Energetics Incorporated, MD
Chris Clark, Senior Scientist, Energetics Incorporated, MD

AIChE 2010 Annual Meeting
Salt Lake City, UT
November 10, 2010

Industry Consumes 1/3 of Energy in U.S.

Reducing U.S. industrial energy intensity is essential to achieving national energy and carbon goals

Petroleum	38.3%
Natural Gas	32.2%
Electricity*	13.8%
Coal and Coke	7.6%
Renewable Energy	8.0%

* Excludes losses

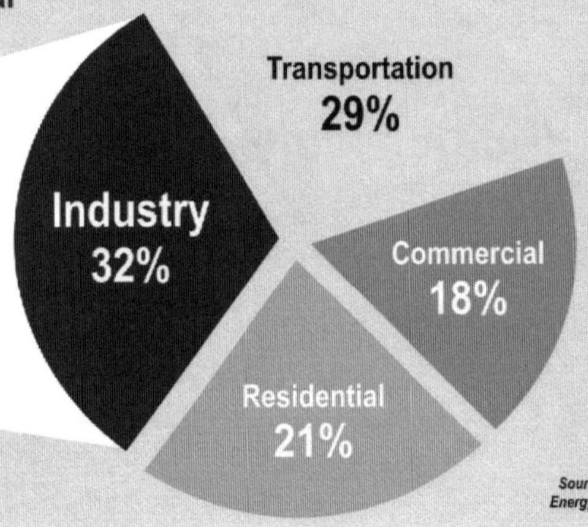

Source: EIA Annual Energy Review 2007.

ITP Objectives

- Reduce industrial energy intensity 25% in 10 years
- Contribute to 2050 goal of 80% reduction in GHG emissions from 2005 levels
- Establish the U.S. as the global leader in energy management

In addition to being the largest energy-using economic sector, U.S. industry…

- Has >200,000 sites
- Spurs job creation and investment
- Produces nearly ¼ of world manufacturing output
- Produces $6+ trillion in goods; $1 trillion in exports
- Is responsible for ~1,660 $MMTCO_2$ per year from energy consumption
- Makes the highest contribution to U.S. GDP (12% from manufacturing alone)

Water Efficiency: A Critical National Priority

- Major U.S. industries are at risk if water shortages occur.
- Energy consumption in most water systems worldwide could be reduced by at least 25% through cost-effective efficiency actions.*
- A 5% reduction in withdrawals could save industry an estimated $330 million/yr and the thermoelectric power industry over $3.5 billion/yr.**
- This workshop with stakeholders from industry, academia, national labs, and government is to identify opportunities for desalination and industrial water efficiency.

Water-Intensive Industries*

#1: Forest products, mainly for *processing*
#2: Steel
#3: Petroleum
#4: Chemicals
(#2–#4: Mainly for *cooling*)

*"Industrial Water Use and Energy Efficiency", Energetics, 2001 & "Water Use in Industries of the Future," CH2M Hill, 2003

1. Desalination opportunities include:

- Waste heat recovery and reuse
- Solar thermal and other renewables
- Low-energy process changes
- Selective membrane enhancements

2. R&D opportunities for low-energy, low water-intensity processes for manufacturing include:

- More efficient steam generation
- Waste heat recovery and reuse
- Materials recycling, including scrap-based metals production
- Innovative pulping/papermaking processes
- Advanced forming and finishing processes with reduced cooling requirements

Operating Costs for Large Thermal Desalination Plant

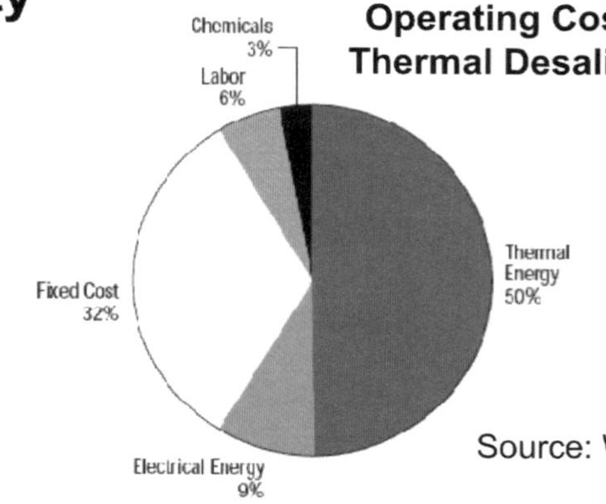

Source: Wangnick 2002

*Alliance to Save Energy
**Internal estimates based on USGA and EPRI data

Meeting Future Industrial Water Demand a Growing Concern

- The United States withdraws an estimated 410 billion gallons of water per day:
  - Self-Supplied Industrial Withdrawals are an estimated 18.2 billion gallons per day.
  - Public Supply Withdrawals, which service industry, are an estimated 44.2 billion gallons per day.

Industrial Withdrawals by source & State, 2005

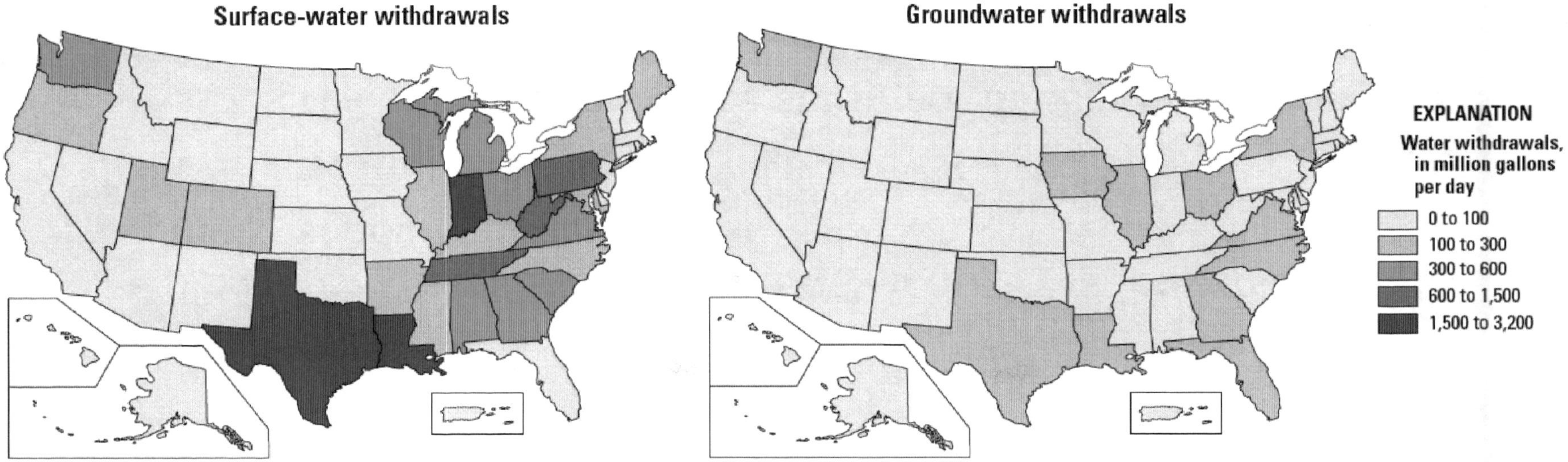

Source: USGS, *Estimate Use of Water in the United States in 2005*

Industrial Water Uses

- **Cooling water**
  - The largest single use of water by industry is for cooling in thermal power generation
- **Process water**
  - Industry uses water to make steam for direct drive power and for use in various production processes or chemical reactions
- **Water for products**
  - E.g., food, beverage and pharmaceutical sectors consume water by using it as an ingredient in finished products for consumption
- **Water as a medium for waste disposal**
  - Many businesses dispose of wastewater or cleaning water into natural fresh water systems., e.g. rivers and lakes
- **Water for energy**
  - Multi-purpose hydro projects manage water for many interests: flood control, irrigation, recreation, drinking water, energy

Water – Energy Nexus

The relationship between energy and water use is complex as every use of water may require more energy.

- Heating
- Cooling
- Treating
- Moving
- Disposing

Each processes requires energy input and adds to the bottom line of facilities expenses

There is Some Recent Policy Attention to the Energy-Water Nexus in the U.S. Congress:

- *Energy and Water Research Integration Act of 2009*
  - Calls for studies and assessments of the energy-water nexus
- *American Clean Energy and Security Act of 2009 (ACES)*
  - Calls for changes to the energy mix with implications for water use

Workshop Goals

- Desalination
  - Identify the barriers and opportunities to improve desalination efficiency
  - Identify R&D pathways to achieve cost effective water purification technologies
  - Develop a vision and roadmap for desalination to meet future industrial demands
- Water use efficiency in manufacturing
  - Identify R&D pathways to reduce water intensity in manufacturing processes
  - Identify manufacturing processes that can cost effectively use lower-quality water without harming process integrity
  - Develop an industry vision and roadmap for low water intensity manufacturing

Workshop Details

- Held on February 23–24, 2010 in Tampa, FL
- Workshop was sponsored by the Industrial Technologies Program (ITP) of the U.S. Department of Energy's Energy Efficiency and Renewable Energy Office
- Stakeholder participation from a cross-section of the public sector, academia, and private industry
- Objectives were to
  - Identify high-impact R&D pathways to address the barriers facing increased desalination technology development and deployment
  - Identify high-impact R&D pathways to overcome barriers to improving target areas in industrial water efficiency in industry
- Results were the development of R&D pathways in specific areas for both desalination technology and improved water use efficiency

High Impact Research and Development Pathways for DesalinationTechnology - I

Development of Energy Efficient and Economical Technology for Zero Liquid Discharge

Basic Science Knowledge	Development/ Applied Research	New Application/ Technology Solution
• Need to fully understand the science of removing constituents from concentrate for beneficial uses • Work can include ionic saturation, super saturation and crystallization • Develop understanding of evaporative cooling	• Develop energy efficient brine concentrators and crystallizers • Improve heat transfer from concentrate • Improve materials for construction • Build cooling towers and thermodynamics • Improve pretreatment through softening and ion exchange • Evaluate more salt tolerant flora and fauna	• Develop more energy efficient and cost effective zero liquid discharge systems • Improve commercial products: road salt, sodium sulfate, calcium, magnesium, and cement additives • Develop corrosion resistant cooling towers for salt removal from concentrate • Improve brine concentration for concentration driven power generation/ion exchange • Develop saline wetlands

High Impact Research and Development Pathways for DesalinationTechnology - II

Development of New "Smart Membranes" and Associated Technology

Basic Science Knowledge	Development/ Applied Research	New Application/ Technology Solution
• Need to fully understand the membrane separation processes • Improve membrane surface chemistry • Improve imbedded sensor and control development and engineering • Advance research of new membrane materials • Integrated technology using thermal, electrical, chemical, biological and membrane separation methods • Understand how membranes fail	• Improve radical membrane module design and optimization • Manage pretreatment based on real-time information from the membrane • Improve energy efficiency using sensors and controls • Improve efficiency of selective properties • Manage fouling film formation	• Need higher process intensification resulting in lower energy requirements for desalination by "working" the membrane surface even more.

High Impact Research and Development Pathways for DesalinationTechnology - III

Biomimicry Applications

Basic Science Knowledge	Development/ Applied Research	New Application/ Technology Solution
• Need a better understanding of ion transfer in biological systems, including biological thermodynamics • Improve understanding of biological membrane separation • Characterize and analyze biochemical reactions/processes on salt/water regulation • Understand how different species manage moisture, minerals, and foreign substances	• Engineer biological process/systems in the laboratory for beneficial use • Research mechanically and chemically durable membrane properties • Use biological principles to enhance selectivity for contaminant or membrane properties • Develop self-healing biological membranes • Conduct research into "self powered" biological separation • Use biomimicry to develop low fouling membranes	• Develop new membranes using biological principles for separation • Engineer biological membranes (e.g. aquaporin) for separation processes • Create biological membranes tailored to address contaminants • Create process that transfer ion from solution to similar electro dialysis • Improve humidification-dehumidification on thermal processes • Develop new products to reduce scaling

High Impact Research and Development Pathways for DesalinationTechnology - IV

Development of a Quantitative Standard Procedure to Predict Membrane Scaling/Fouling

Basic Science Knowledge	Development/ Applied Research	New Application/ Technology Solution
• Need better understanding of the chemistry associated with the scaling of membranes, including fundamental knowledge of particles characterization, microbiology, and analytical and aquatic chemistry	• Develop a new analytical protocol • Develop benchmarks • Conduct bench scale and pilot plants tests • Validate test procedures • Develop new instruments	• Create a software/program for incorporation in plant control and monitoring systems • Develop new industry standards

High Impact Research and Development Pathways for DesalinationTechnology - V

Techno-Economic Model to Evaluate Integrated Power and Water Desalination Plants

Basic Science Knowledge	Development/ Applied Research	New Application/ Technology Solution
• Improve unit operations and processes • Develop thermodynamics • Better understand economics of plant citing and operations • Improve energy analysis • Study heat transfers and corrosion resistant chemistry	• Improve process modeling • Create cost database of capital equipment to enable rolled-up economic models • Explore systems engineering and applied thermo-economic analysis • Study efficient corrosion resistant heat exchange methods	• Better understand hybrid water-energy systems • Improve cost optimization tools and current databases • Develop system design and economic modeling tools • Create economically viable, distributed (small-scale)/power-water systems

High Impact Research and Development Pathways for DesalinationTechnology - VI

Membranes for Forward Osmosis

Basic Science Knowledge	Development/ Applied Research	New Application/ Technology Solution
• Develop better understanding of forward osmosis membrane structure, chemistry, and biology	• Develop new and improved membrane materials and configurations	• Improve economics and develop a full-scale application

High Impact Research and Development Pathways for DesalinationTechnology - VII

Separation Process Technology

Basic Science Knowledge	Development/ Applied Research	New Application/ Technology Solution
• Need a better understanding separation processes • Develop a better understanding of how acoustic and electrical fields may remove contaminants form solution • Develop understanding of how biological adsorption and decomposition may play a role in the removal of contaminants	• Improve proof of concept ideas by bench-marking and pilot scale testing • Optimize the configuration, design and operation of key technologies • Identify applications and challenges (interfaces, fouling, and scaling) • Conduct a techno-economic analysis	• Develop new selection of contaminant removal options

High Impact Research and Development Pathways - Industrial Water Use Efficiency

Eight strategic targets identified to improve industrial water efficiency

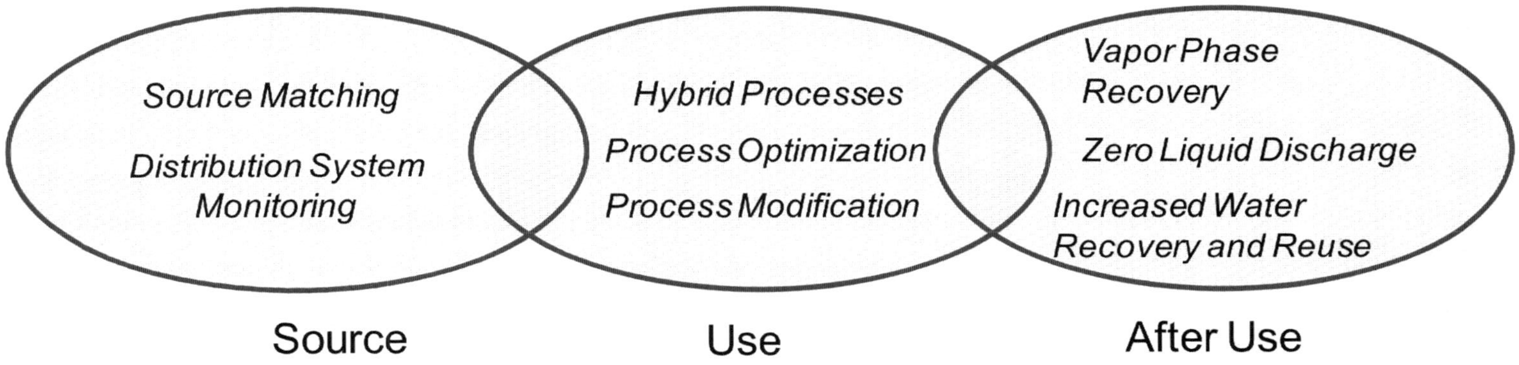

High Impact Research and Development Pathways - Industrial Water Use Efficiency - I

Reducing Water Losses through Improved Flow Management

Basic Science Knowledge	Development/Applied Research	New Application/Technology Solution
• Understanding physics of water (e.g. dipole) movement in confined space (e.g. pipe)	• Development of noninvasive flow rate measurement technology	• Cheap flow meters that can be used with or calibrated to a variety of liquids
• Understanding propagation of ultrasonic signal through clean water and potential Doppler frequency shifts	• Ultrasonic Doppler flow meters or similar technology for measurement of clean water flow rates	• Relatively cheap Doppler flow meter that sacrifices sensitivity for precision

High Impact Research and Development Pathways - Industrial Water Use Efficiency - II

Vapor and Energy Recovery

Basic Science Knowledge	Development/Applied Research	New Application/Technology Solution

• Advance understanding of the following concepts:
  • Fluid flow
  • Heat transfer
  • Materials science
  • Water absorption
	• Research and develop vapor separation membranes (polymeric, ceramic, metallic)
	• Research and develop capillary condensation membranes
	• Develop or improve methods to use recovered energy (e.g., to preheat boiler feed water)
	• Improve efficiency in turbine generation
	• Improve efficiency of heat exchangers
		• Improved energy and vapor recovery throughout systems in the following industries:
		  • Pulp and paper
		  • Textiles
		  • Food drying
		  • Power plant cooling
		  • Metals processing
		  • Wet scrubbers

High Impact Research and Development Pathways - Industrial Water Use Efficiency - III

Membrane Materials

Basic Science Knowledge	Development/Applied Research	New Application/Technology Solution

• Improve integration of materials science with water chemistry
• Improve understanding of nanomaterials and nanoengineering
• Improve understanding of non-polymer-based materials (e.g., ceramics)
• Develop nanoscale hydraulics modeling
	• Develop membranes with the following characteristics:
	  • Pressure tolerant
	  • Temperature resistant
	  • Chemical resistant
	  • Fouling resistant
	  • Corrosion resistant
	  • Scaling resistant
	  • Improved ion selectivity
		• Durable membranes with high flux capacity which are cost-effective, requiring low capital and operations and maintenance costs

High Impact Research and Development Pathways - Industrial Water Use Efficiency - IV

Water Separation Chemistry

Basic Science Knowledge	Development/Applied Research	New Application/Technology Solution
• Improve understanding of chemical properties of process water	• Develop separation techniques tailored to harsh environments to remove the following components: • Bad components for disposal • Good components for further use	• Broader range of process water quality that is usable • Improved discharge water quality and public health
• Improve understanding of nanomaterials and control in situ behavior	• Develop techniques to target selective ions for removal	• Improved recycling and reuse of discharge stream • Enhanced process efficiency and reduced energy consumption

High Impact Research and Development Pathways - Industrial Water Use Efficiency - V

Contaminant Removal and Zero Liquid Discharge Sludge

Basic Science Knowledge	Development/Applied Research	New Application/Technology Solution
• Develop understanding of dissolved organic carbon (DOC) and its interaction with the environment	• Improve tools to characterize DOC • Develop processes to remove DOC • Explore mechanisms that result in endocrine removal • Exploring biological processes to remove DOC	• Direct and indirect water reuse • Enhanced water recycling • High potential for use with nanofiltration
• Improve characterization of specific organic compounds • Develop understanding of physics behind plasma treatment	• Develop tools and techniques to treat industrial sludge and bio-solids	• Use sludge as Fuels and/or Feedstock • Land applications
• Improve understanding of water quality and chemistry	• Develop tools and techniques to achieve and implement zero liquid discharge	• Distributed water systems and water cascading • Internal water recycling

Summary

The Industrial Technology Program at DOE will continue to engage in existing R&D efforts as well as in new research projects to evaluate the most suitable technologies and processes for obtaining and processing both potable and non-potable water resources.

For additional information please contact:
bhima.sastri@ee.doe.gov

(494a) Carbon Dioxide Capture Using Ionic Liquid 1-Butyl-3-Methylimidazolium Acetate

Mark Shiflett[1], David Drew[2], Robert Cantini[2] and A. Yokozeki[3], (1)DuPont Central Research and Development, Wilmington, DE, (2)DuPont Engineering, Wilmington, DE, (3)DuPont Fluorochemicals, Wilmington, DE

Carbon dioxide capture using aqueous amine scrubbing is currently considered the most feasible option for separating CO2 from post-combustion flue gas. Using simple absorption and stripping configurations, monoethanolamine has been commercially demonstrated to effectively scrub CO2 from post-combustion flue gas. However, the current capital and operating costs are high and do not meet the Department of Energy's target to remove 90% of the CO2 from post-combustion flue gas with no more than a 35% increase in the cost of electricity. The evaluation of advanced absorbents, adsorbents and membranes is under way in order to find the most energy-efficient CO2 capture technology. We have modeled an ionic liquid that can reduce the energy losses by 16% compared with a commercial monoethanolamine process. The choice of the ionic liquid, 1-butyl-3-methylimidazolium acetate, has not been optimized but was chosen based on chemical absorption behavior and the desire to understand performance. Engineering design estimates indicate that the investment for the ionic liquid process will be 11% lower than the amine-based process and provide a 12% reduction in equipment footprint. A parametric study indicates that the energy losses may be reduced by an additional 33%, which could meet the Department of Energy CO2 capture target.

See more of this Session: Separations Needs for CO2 Capture I
See more of this Group/Topical: Topical 1: Separation Needs for Energy Independence and Environmental Sustainability

(494b) CO_2-Philic Oligomers as Alternatives for PEGDME in CO_2 Absorption

Matthew B. Miller, Dept.of Chemical and Petroleum Engineering, University of Pittsburgh, Pittsburgh, PA, Robert M. Enick, Department of Chemical and petroleum Engineering, University of Pittsburgh, Pittsburgh, PA and David Luebke, US DOE/NETL, Pittsburgh, PA

Advanced gasification power plants will employ the water-gas shift reaction producing a high pressure gas-phase mixture containing CO_2, H_2 and water. A polydisperse mixture of poly(ethylene glycol) di-methylethers, PEGDME, can be used as a solvent for the absorption of CO_2. The objective of this study was to identify alternative physical solvents that would selectively dissolve only CO_2 from this mixture, while exhibiting low viscosity.

The first phase of our research involved the comparison of the CO_2 solvent strength of PEGDME with that of other low volatility oligomers that are known to be "CO_2-philic". These oligomeric solvent candidates include poly(propylene glycol) di-methyl ether (PPGDME), poly(propylene glycol) di-acetate (PPGDAc), poly(butylene glycol) di-acetate (PBGDAc) with linear C4 monomers, poly (dimethyl siloxane) (PDMS), perfluoropolyether (PFPE), and glyceryl tri-acetate (GTA). Pressure-composition phase diagrams are presented for the pseudo-binary systems of CO_2 with PEGDME n=6, PPGDME n=6, PDMS n=6, PBGDAc n=3.2, PPGDAc n=6.7, and GTA (which is analogous to a trimer of polyvinyl acetate), each at 25C and 40C. Henry's law constants were also calculated for each pseudo-binary system in the high polymer concentration region. Although the performance of PPGDME, PEGDME and PDMS are comparable on a weight basis, PPGDME and PDMS appear to be the best CO_2 solvents based on the ability to absorb CO_2. Further, the viscosity of these compounds at 22C and 40C indicates that PDMS is significantly less viscous than all of the others, including PEGDME. Further, PDMS and PFPE are essentially immiscible with water, and water is only slightly soluble in PPG- and PBG-based oligomers and GTA, whereas water and PEGDME are completely miscible.

Small volatile CO_2 solvents were also examined in the second phase of this work. Phase behavior results in the form of pressure-composition diagrams are presented for the binary systems of CO_2 and N-formylmopholine, methanol propylene carbonate, acetone, 1,4 dioxane, isooctane, methyl acetate, 2-butoxy ethoxy acetate, and 2-(2-butoxyethoxy) ethyl acetate. Acetone and 1,4-dioxane are the most CO_2-philic compounds however, their high vapor pressures prevent them from being viable solvents.

See more of this Session: Separations Needs for CO2 Capture I
See more of this Group/Topical: Topical 1: Separation Needs for Energy Independence and Environmental Sustainability

(494c) Absorption of CO2 Into Aqueous Potassium Carbonate Solution Promoted by Carbonic Anhydrase Enzyme for Post-Combustion CO2 Capture

Yongqi Lu[1], Xinhuai Ye[1], Massoud Rostam-Abadi[2] and Robert Patton[3], (1)Advanced Energy Technology Initiative, University of Illinois at Urbana-Champaign, Champaign, IL, (2)Advanced Energy Technology Initiative; Department of Civil & Environmental Engineering, University of Illinois at Urbana-Champaign, Champaign, IL, (3)US DOE/ National Energy Technology Laboratory, Pittsburgh, PA

A novel Integrated Vacuum Carbonate Absorption Process (IVCAP) for post-combustion CO_2 capture is being developed by the Advanced Energy Technology Initiative of the Institute of Natural Resource Sustainability at the University of Illinois at Urbana-Champaign. The solvent used in the process is potassium carbonate (PC). A unique feature of this process is its ability to use either waste or low quality steam from a power plant's low pressure turbine for CO_2 desorption. As a result, electricity loss due to steam extraction is significantly lower than that of mono-ethanol-amine (MEA) processes. However, a key technical issue for the IVCAP is the slow absorption rate of CO_2 into the PC solution. In this study, the *carbonic anhydrase* (CA) enzyme was experimentally evaluated as a biocatalyst to promote the rate of CO_2 absorption into PC solutions.

The activity of the CA enzyme and its long-term chemical and thermal stability were evaluated using a stirred cell reactor under various conditions of temperature, enzyme dosage level, concentration of PC solution, initial CO_2 loading in the solution, CO_2 partial pressure, and concentrations of various contaminants. It was found that the rate of CO_2 absorption was promoted by 3-15 times at a dosage level of 300 mg/L CA between 25°C and 50°C. The CA enzyme exhibited a stable activity over a two-month period. Activity tests over a longer time are in progress. The effect of typical impurities in a flue gas on the enzyme activity was simulated by exposing the CA enzyme to a solution containing SO_4^{2-} (0.1-0.9M), NO_3^{2-} (0.05-0.2M) and Cl^- (0.3-0.7M), either individually or in a mixture. It was revealed that the enzyme lost only less than 11% of its initially activity.

Process engineering data collected to date indicate that the CA enzyme is a promising catalyst for promoting the rate of CO_2 absorption into the PC solution. Ongoing research is focusing on testing several CA enzymes obtained from different sources, developing a CA immobilization method, and evaluating the potential application of the IVCAP for simultaneous removal of SO_2 and capture of CO_2 from coal combustion flue gases.

See more of this Session: Separations Needs for CO2 Capture I
See more of this Group/Topical: Topical 1: Separation Needs for Energy Independence and Environmental Sustainability

(494d) Physico-Chemical Properties of Functionalized Ionic Liquid Solutions for Kinetic Studies of CO2 Absorption

Burcu Gurkan, Mark J. McCready and Joan F. Brennecke, Chemical and Biomolecular Engineering, University of Notre Dame, Notre Dame, IN

Ionic liquids (ILs) are receiving growing attention as liquid absorbents alternatives to aqueous amine solutions for the removal of CO2 from flue gas in post-combustion processes, due mainly to their thermal stability and nonvolatility. We have synthesized tetraalkylphosphonium based ionic liquids and have shown that the CO2 uptake capacity can be increased up to 1 mol of CO2 per mole of IL by appending the amine functionality to the anion[¥]. While the high CO2 capacity is desired, the relationship between the equilibrium and kinetics of the absorption reaction is important to understand when designing the absorber-stripper process.

Our preliminary studies based on water-miscible ILs show that they can have significantly faster reaction rates than aqueous amines. Second order reaction rate constants are estimated to be up to one order of magnitude higher than monoethanol amine. However not all functionalized ILs are miscible with water. Furthermore water is not desired as a reactive dilution medium, due to its high heat capacity. Therefore functionalized ILs are mixed with inert, low-viscosity and thermally stable solvents that have negligible vapor pressure at the measurement conditions. Physical properties such as the density and viscosity of the solutions as well as the solubility and diffusivity of CO2 in these solutions are required to study the kinetic properties. The density and viscosity of possible IL solutions are measured in the temperature range of 22 to 80 C and in the IL concentration range of 0.05 to 1 mol/L. The physical solubility of CO2 in these solutions is measured in-situ with an ATR-FTIR probe. Diffusivity of CO2 is most important for ILs when they are not diluted because the process can become mass-transfer limited owing to their high viscosities. For the IL solutions, diffusivities are estimated using a modified Stokes-Einstein relation. These physico-chemical properties and relationships enable the kinetic studies of CO2 absorption in ILs.

[¥] B. E. Gurkan, J. C. de la Fuente, E. M. Mindrup, L. E. Ficke, B. F. Goodrich, E. A. Price, W. F. Schneider and J. F. Brennecke, Journal of the American Chemical Society, 2010, 132, 2116-+.

See more of this Session: Separations Needs for CO2 Capture I
See more of this Group/Topical: Topical 1: Separation Needs for Energy Independence and Environmental Sustainability

(494e) Thermal Stability of Ionic Liquids Studied by TGA, DSC and GC/MS

Yong Huang and Joan F. Brennecke, Chemical and Biomolecular Engineering, University of Notre Dame, Notre Dame, IN

Ionic liquids (ILs) are organic salts with low melting points (below 100°C). They are being studied for a variety of applications, including as solvents for reactions, as high temperature heat transfer fluids, and as solvents to separate CO_2 from flue gas. Thermal stability and decomposition are important properties, especially for applications with high temperature environments. This work focuses on the thermal stability of ionic liquids with phosphonium, imidazolium and pyridinium cations. Temperature ramped and isothermal experiments were performed using thermal gravimetric analysis (TGA). Glass transitions (T_g), cold crystallization temperatures (T_{cc}), melting points (T_m) and heat capacities were measured by differential scanning calorimetry (DSC). The decomposition products were analyzed by gas chromatography/mass spectrometry (GC/MS).

Dynamic and static decomposition experiments of ILs were performed with a Mettler Toledo TGA. The decomposition temperatures under two different gaseous environments (nitrogen and air to represent inert and oxidizing environments) were evaluated. For dynamic decomposition, one important descriptor of thermal stability is the decomposition onset temperature (T_{onset}) at a temperature ramp rate of 10 °C/min under a nitrogen environment. For the isothermal decomposition, all experiments were performed at temperature above 100 °C for 16 hours under nitrogen or air, the data was evaluated by using a pseudo zero-order reaction mechanism because the weight vs. time curve is linear, and then activation energy and $T_{1\%/1day}$ (the temperature at which 1% weight loss occurs in 1 day) were derived.

Glass transitions (T_g), cold crystallization temperature (T_{cc}), and melting point (T_m) were measured by DSC from Mettler Toledo. Many ILs tend to subcool easily to form glasses at very low temperatures rather than exhibit crystallization or melting transitions. T_g and T_m values are important for determining the lower limit of operating range where the IL is liquid, and T_{onset} can be regarded as the upper operating range since IL do not evaporate. Heat capacities can be important for evaluating ILs in heat transfer and thermal storage applications.

The decomposition mechanisms under an inert environment were investigated with a VARIAN Single Quadrupole GC/MS, including electron ionization (EI) and chemical ionization (CI). A chromatoprobe and temperature-programmed 1079 injector were used to introduce solid or large volume liquid samples. For tetra-alkylphosphonium ionic liquids, the two primary decomposition pathways involved are: 1) nucleophilic substitution reaction at the α carbon where a nucleophilic anion displaces a trialkylphosphine group and 2) β-elimination where the β-proton is abstracted by a base in concert with the expulsion of trialkylphosphine (Xie et al. Chem. Mater., Vol. 14, No. 11, 2002). For pyridinium and imidazolium ionic liquid, nucleophilic substitution reaction at alkyl group on the nitrogen atoms is the most likely mechanism for the pyrolysis process (W.H. Awad et al. Thermochimica Acta 409 (2004) 3–11).

See more of this Session: Separations Needs for CO2 Capture I
See more of this Group/Topical: Topical 1: Separation Needs for Energy Independence and Environmental Sustainability

Hyperbranched Polymers as Absorbents for CO_2 Capture: Experimental Solubilities and Process Simulation

Walter Martini, Harvey Arellano-Garcia and Günter Wozny

Chair of Process Dynamics and Operation, Berlin Institute of Technology, Berlin, Germany

Keywords: Absorption, Carbon Capture, Hyperbranched Polymers

Introduction

Oxidative coupling of methane (OCM) is an approach for the direct catalytic conversion of methane to higher hydrocarbons, which is reaching widespread attraction among various research groups in the last decade. The conversion to ethylene is the major goal of the process. OCM is a surface induced gas phase reaction and its overall yield is still limited up to 30%. Besides new catalysts, a concept for an integrated downstream process is necessary to overcome this limitation [1].

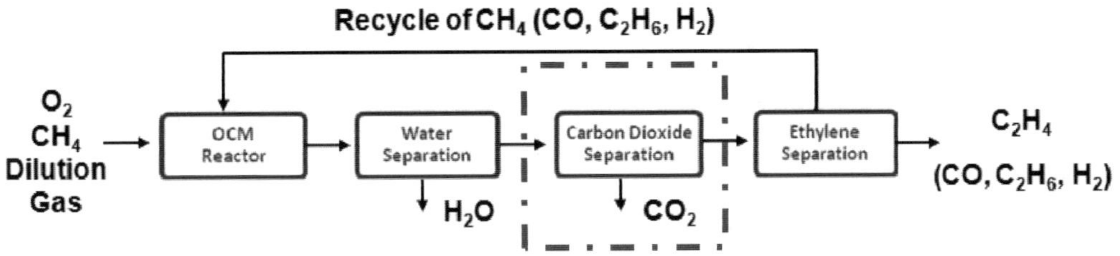

Figure 1. Process flow diagram of the OCM process

Various processes for the OCM have been proposed so far, such as the OXCO process, the UCC process, the ARCO process, the Suzuki process, the Turek-Schwittay process or the Co-Generation process [2]. All of them have in common that the product separation under high pressure and the recycling of unreacted methane is of most importance for the process economics [3]. Because of its current low yield and challenges for an efficient downstream process, OCM has not been applied in the industry yet. Therefore, a miniplant has been built at the Chair of Process Dynamics and Operation of the Technical University of Berlin, where fundamental studies of process alternatives and the effect of recycles and efficiencies of each process unit are being investigated [4]. Figure 2 shows the generalized layout of the miniplant.

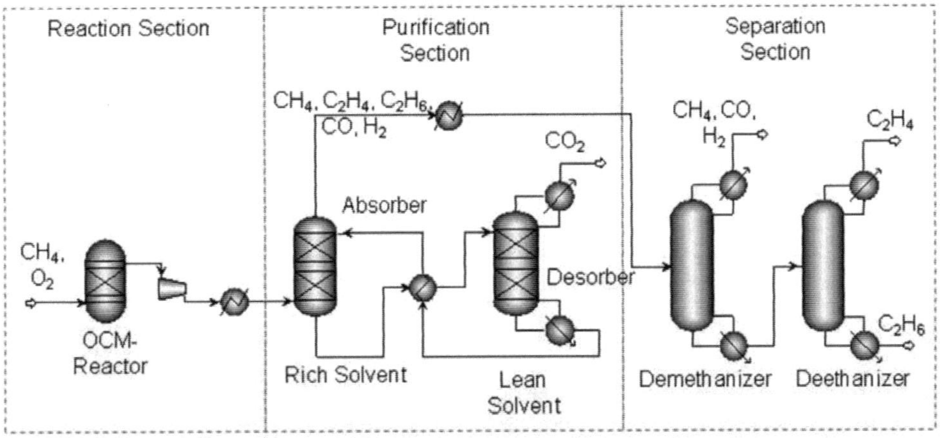

Figure 2. Simplified process flow diagram of the OCM

For process synthesis, the whole OCM process is divided into three general sections: subsequent to the reaction, a purification step takes place, where carbon dioxide has to be removed from the raw gas completely. Finally, in the separation section, the desired product, ethylene, is being separated from all other components, which may be recycled to the reaction section again. Currently, there is the process concept of using carbon dioxide as an inert gas for dilution in the OCM reactor. Thus, the carbon dioxide removal is an even more important step of the downstream process, and therefore, is being focused on.

Different separation techniques have been evaluated for the miniplant, such as adsorption, membranes, absorption and cryogenic separation techniques. Hereby, absorption was found to be most suitable for the given task. Absorption of carbon dioxide is well developed and industrial available. Physical absorption processes like the UOP Selexol® or the Lurgi Rectisol® processes are known, which use dimethyl ether and cold methanol, respectively. Due to a nearly similar solubility of the product and the carbon dioxide in the liquid, those physical absorption processes cause high product losses of more than 30 vol%. Therefore, only chemical solvents like Monoethanolamine (MEA), Diethanolamine (DEA) and Methyldiethanolamine (MDEA) or a mixture of them have been considered so far in the OCM miniplant. These chemicals are used in amine scrubbing processes like the aMDEA® process in different concentration ranges [5]. Standalone rigorous simulations for the absorption process of the miniplant were carried out in Aspen Plus®. As scrubbing liquids, 15 wt% MEA and 30 wt% MDEA solution were compared, where neither MEA nor MDEA as solvent could remove the carbon dioxide totally in a standalone absorption process for the given conditions. As a result, a hybrid membrane amine process for the removal of carbon dioxide in the product stream was proposed, where a two stage membrane system has been combined with an absorption column to form a hybrid process [6].

In this work, hyperbranched polymers are being investigated as an alternative for absorbents for carbon dioxide, since they have shown being promising candidates for gas absorbents with a high capacity for CO_2 and with large selectivities [7].

Hyperbranched polymers

Currently, four major classes of polymers can be distinguished in accordance with their properties and polymeric architecture. Dendritic polymers herein represent the fourth class and are highly branched globular macromolecules. They can be subdivided into three subsets that are related to the degree of structural control, i.e. random hyperbranched polymers, dendrigraft polymers, and dendrimers [8].

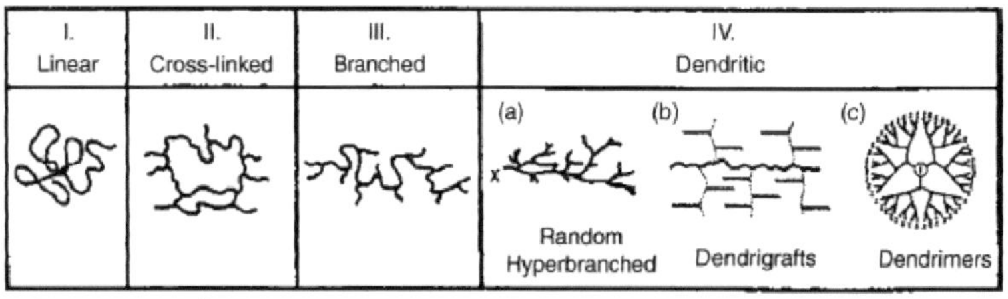

Figure 3. Classes of macromolecules [8]

Dendrimers and hyperbranched polymers have attracted considerable and increasing interest during recent years. They resemble each other in that they are both polymerized from monomers with mixed reactivities, giving branched structures with exponential growth, in both end-group functionalities and molecular weights. Their properties differ strongly from those of linear polymers of the same molar mass (less flexibility, lower entanglement degree, a significant chain-end effect, lower viscosity in solution and in the molten state, high solubility in common solvents, a different relationship between hydrodynamic volume and molar mass, and a different origin of the glass transition temperature) [9].

Dendrimers show a well defined, monodisperse, perfectly branched structure and have a large number of functional end groups [10]. Their generation requires absolute control of all synthesis steps and makes large-scale production difficult and hence expensive. However, many applications do not require structural perfection. Therefore, using hyperbranched polymers can circumvent this major drawback of dendrimers [11].

Unlike dendrimers, randomly branched hyperbranched polymers with similar properties can be easily synthesized via one-step reactions and therefore represent economically promising products also for large-scale industrial applications. At room temperature, many branched polymers exhibit low viscosities in the pure state as well as in solution due to the absence of chain entanglement. Potential applications of hyperbranched polymers range from the use as selective solvents in distillation of azeotropic mixtures or extraction to the control of flow characteristics, as well as the use as drug delivery system and many more.

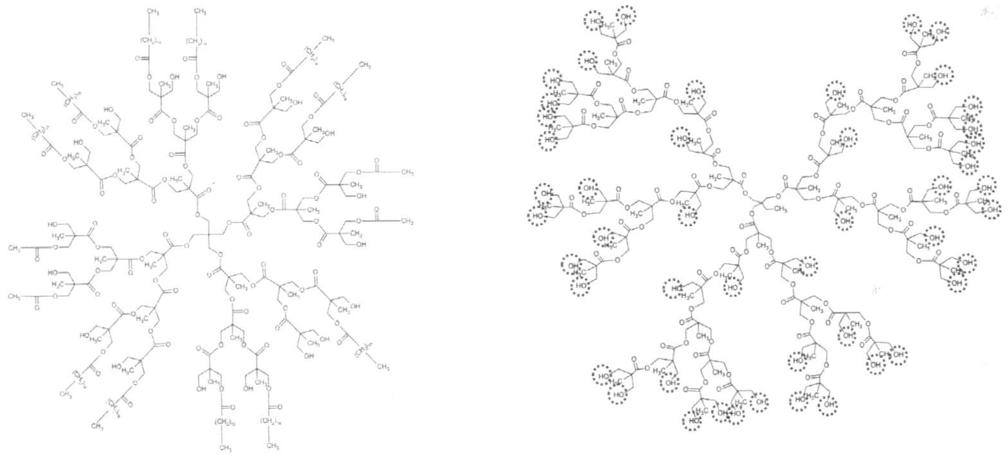

Figure 4. Boltorn® U3000 (*left*) and P500 (*right*) molecules by Perstorp AB (Sweden)

In this work, the hyperbranched polyester/ether polyol Boltorn® P500 and the fatty acid modified dendritic polymer Boltorn® U3000 by Perstorp are introduced to carbon dioxide absorption. Their molecular structure is shown in Figure 4. Boltorn® P500 was chosen because of its high average hydroxyl functionality and, compared to similar polymers, low viscosity. The dendritic polyethylenimine Lupasol FG by BASF SE was investigated as a third solvent. Still, this has to be considered as a first step followed by more detailed investigations on different commercially available polymers and also on tailor-made polymers, which could be optimized in terms of capacity, selectivity, and stability with regard to economic demands.

Measurements and Calculation

Experimental Setup and Procedure

Figure 5 shows a schematic diagram of the experimental setup that was used in this work. A maximum of two equilibrium cells (4) can be measured simultaneously, each having a volume of 80ml. They are provided with a magnetic stirrer and placed inside a water thermostat (5) with a temperature range of 283-363 K. The cell temperatures are determined with a resistance thermometer (Pt 100) in the water bath. The cells are connected via a valve and a quick connection to the gas systems (2) (each 135ml) which contain the solute and each are connected to a pressure sensor. The maximum pressure that can be applied is up to 1.2 bar. The temperature within the gas systems is measured with a resistance thermometer (Pt 100). The gas is fed from a reservoir (1) to the system. The cells and the gas systems are attached to a vacuum pump (6) for evacuation and desorption.

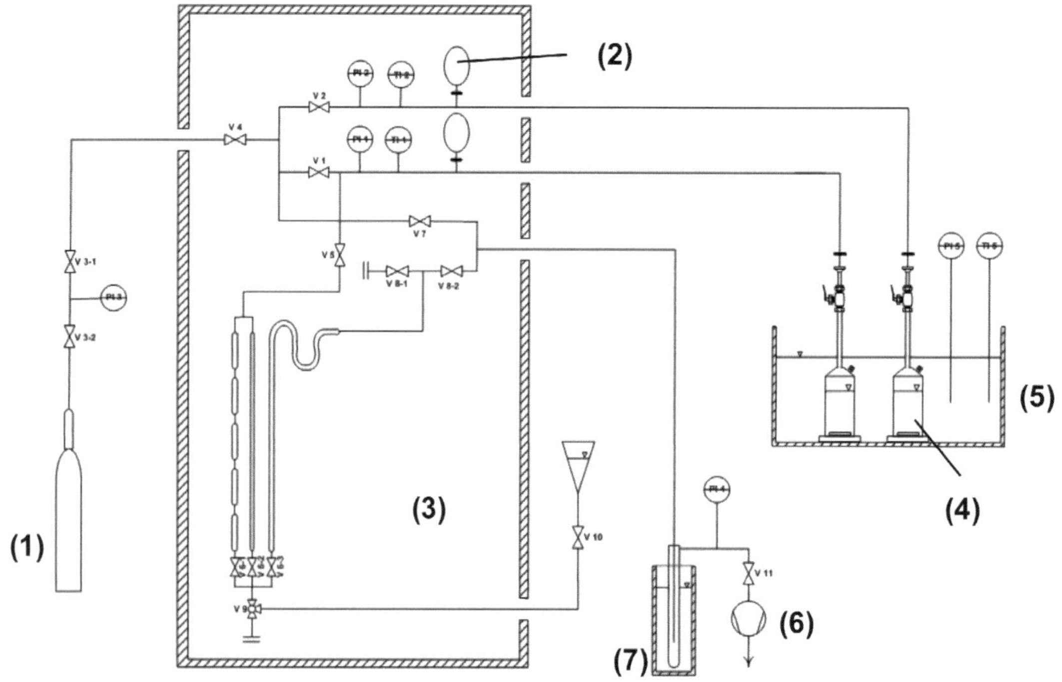

Figure 5. Schematic diagram of the gas solubility apparatus: (1) CO_2 supply; (2) gas reservoir; (3) isothermal environment; (4) equilibrium cell with magnetic stirrer; (5) water thermostat; (6) vacuum pump; (7) cooling trap;

The degassed solvent (liquid polymer or polymer solution) is filled into the equilibrium cell at constant temperature and stirred continuously. After evacuating the system and the cell, the valve between the cell and the system is closed and the gas is fed from the reservoir (1) into the gas system. Then, the gas is admitted into the cell and absorbed until equilibrium is reached (i.e. a constant pressure is observed). Due to usually higher solubilities at lower temperatures, it is best to start with the highest temperature and decrease it in the course of the experiments. The time of equilibration of the whole system (system + cell) depends on the viscosity of the solvent and the solubility and takes roughly 90 minutes. For reproducibility, experiments were repeated three times with two equilibrium cells.

Densities of the polymer and its solutions with ethanol were measured in the corresponding temperatures range using a pycnometer with a volume of 25ml. A Haake rotational viscometer was used for the measurement of dynamic viscosities in the range of 2–10000 mPas.

Calculation of the Henry coefficients

Boltorn® U3000 and Lupasol FG could be investigated as pure liquids, but Boltorn® P500 was mixed with ethanol (resulting solution: 33% weight fraction of ethanol) in order to decrease the viscosity of the solvent. Assuming that no polymer is in the gas phase, the amount of absorbed CO_2 could be determined iteratively by PVT-calculations of the gas system and the gas in the cell, and subsequently solving mass balances:

$$n_{CO_2}^{abs} = n_{CO_2,0}^{gas\,system} - n_{CO_2,l}^{gas\,system} - \left(\frac{V^{cell} - \dfrac{m_{ethanol} + m_{polymer} + \widetilde{M}_{CO_2} \cdot n_{CO_2,cell}^{abs}}{\rho_{CO_2+polymer+ethanol}^{L}}}{\upsilon_{CO_2+ethanol,cell}^{G}} \right) \quad (1)$$

The molar amount of CO_2 in the gas system as well as the molar volume of the gas phase in the cell were calculated using the Soave-Redlich-Kwong equation of state (SRK). The density of the liquid phase was found to be almost independent of the amount of CO_2 absorbed. Therefore, only the density of the pure solvent mixture is considered. The mass-fraction was then easily achieved by:

$$w_{CO_2}^{abs} = \left(\frac{\widetilde{M}_{CO_2} \cdot n_{CO_2,cell}^{abs}}{\widetilde{M}_{CO_2} \cdot n_{CO_2,cell}^{abs} + m_{ethanol} + m_{polymer}} \right) \quad (2)$$

Since the mass-fraction ($w_{CO_2} < 0.05$) and the pressure ($p < 800mbar$) were low, the Poynting correction and activity coefficient could be neglected [12] such that the mass-fraction based Henry coefficient could be calculated by

$$H_{CO_2,polymer} = \frac{y_{CO_2} \cdot p}{w_{CO_2}^{abs}} \quad (3)$$

Results and conclusions

Figure 6 shows the results of the viscosity measurements as a basis for their applicability in an industrial process.

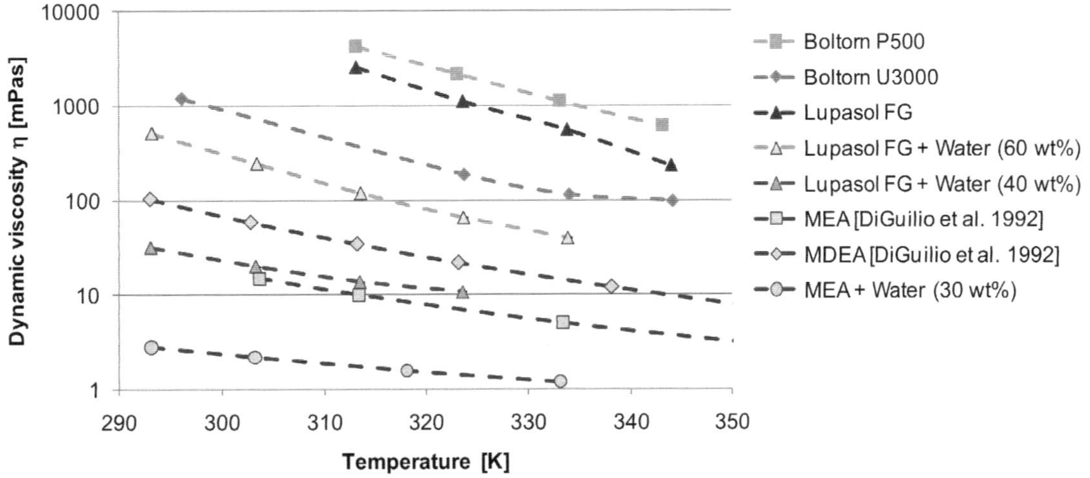

Figure 6. Dynamic viscosities of polymers and polymer solutions

As can be seen, Boltorn U3000 has a high viscosity which gets into a more feasible region at higher temperatures. Boltorn P500 shows very high viscosities even at a higher temperature range, which was the reason for mixing it with ethanol, while Lupasol FG was investigated both as a pure solvent as well as mixed with water.

In Figures 7a and 7b, the calculated mass-fraction based Henry coefficients of CO_2 in the respective solvent can be seen.

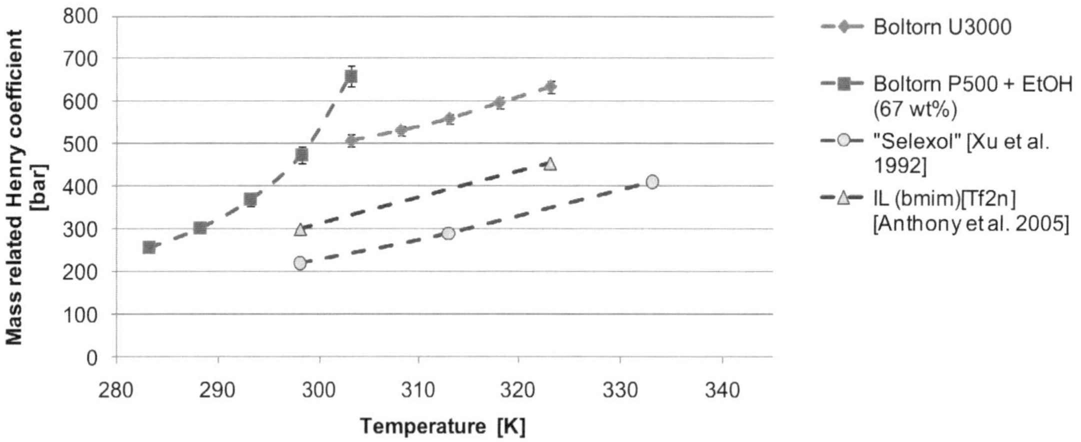

Figure 7a. Mass-fraction based Henry coefficients of CO_2 in different polymers

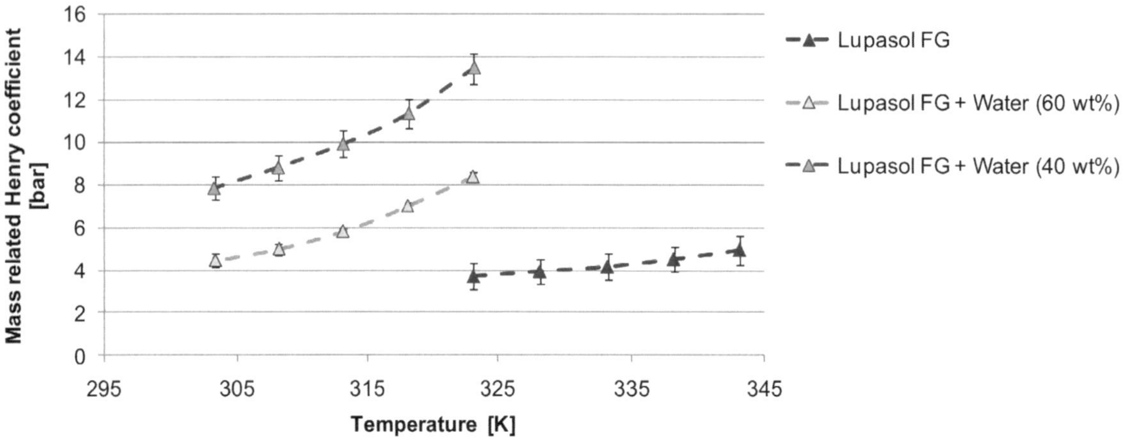

Figure 7b. Mass-fraction based Henry coefficients of CO_2 in Lupasol FG and its mixtures

In addition to the thermodynamic experiments, kinetic investigations on the CO_2 absorption behavior in hyperbranched polymers are still on the way. The kinetics play also an important role in absorption but are not discussed in this work.

Moreover, simulation studies using Aspen® will be presented to validate the potential of hyperbranched polymers in removing CO_2. Furthermore, a miniplant is being built for the whole OCM downstreaming process within the UniCat project [13]. Thus, suitable hyperbranched polymers will be investigated further on a larger scale in the downstreaming miniplant.

Acknowledgements

The authors acknowledge support from the Cluster of Excellence "Unifying Concepts in Catalysis" coordinated by the Technische Universität Berlin and funded by the German Research Foundation - Deutsche Forschungsgemeinschaft. We would also like to thank Perstorp AB and BASF SE for providing us with samples of different hyperbranched polymers.

References

[1] M. Driess et.al., *Evaluation Presentation of the Cluster of Excellence for the German Research Foundation*, Bad Honef, (2007)

[2] E.E. Wolf, *Methane conversion by oxidative processes,* Van Nostrand Reinhold (1992)

[3] D. Salerno et al., *Separación de Etileno Mediante Alimentación* Dividida a Partir de Gases Livianos, CAIP '2009, Montevideo (2009)

[4] S. Stünkel et al., *Entwicklung eines energieeffizienten Hybridprozesses für die CO_2-Abtrennung*, Chemie Ingenieur *Technik* 82(2010) 1416-1417

[5] D. Salerno et.al., *Computer Aided Chemical Engineering*, 27(2009), 855-860

[6] Stünkel et.al., *17th International Conference on Process Engineering and Chemical Plant Design*, (2008), Cracow, Poland

[7] J. Rolker et.al., *Ind. Eng. Chem. Res.*, 46(2007) 6572-6583

[8] D.A. Tomalia and J.M.J. Fréchet, *Journal of Polymer Science: Part A: Polymer Chemistry*, 40(2002) 2719-2728

[9] E. Zagar and M. Zigon, *Macromolecules*, 35(2002) 9913-9925

[10] D.A. Tomalia et.al., *Polymer Journal*, 17(1985) 117-132

[11] M. Seiler, *Fluid Phase Equilibria*, 241(2006) 155-174

[12] J.J. Carrol, *Chemical Engineering Progress*, 10(1991) 48-52

[13] Unifying Concepts in Catalysis, *http://www.unicat.tu-berlin.de/*

(507a) Characterizations of Surface Nano-Structured RO/NF Membranes: Surface Properties and Membrane Performance

Kari J. Varin, Nancy H. Lin and Yoram Cohen, Chemical and Biomolecular Engineering, University of California, Los Angeles, Los Angeles, CA

Improving the performance of current RO/NF membranes requires significant reduction in membrane fouling and mineral scaling propensities. Previous strategies for mitigating membrane fouling via surface modification have relied on alteration of the membrane surface chemistry and topography by addition of a permselective polymer thin film, coated or graft polymerized, that would act both as a separation layer and a physical boundary to prevent adsorption of foulants. In the present study, a novel atmospheric pressure plasma-induced graft polymerization (APPIGP) method was developed to enable the generation of a high surface density of active surface sites on a polyamide membrane for subsequent graft polymerization using a suitable monomer. Surface structuring via graft polymerization was then carried out, using three different water soluble monomers, to form a brush layer of polymeric chains that are covalently and terminally bound to the surface. The chemical and physical features of the resulting grafted polymer films were tuned by the selection of monomer chemistry as well as the reaction conditions to achieve unique surface architectures to reduce surface fouling while achieving the desired rejection and high water permeability. The presence of grafted polymeric brush layer was elucidated by Fourier Transform Infrared (FTIR) Spectroscopy. Graft polymerization kinetics were measured with respect to the change in grafted polymer layer thickness with reaction time and monomer concentration, and the change in membrane surface energy was determined based on contact angle measurements. Furthermore, the surface nano-structured RO/NF membrane performance was evaluated by examining the membrane mineral salt scaling and biofouling propensities. Membranes fouling propensity was assessed via flux decline studies using model proteins and polysaccharides solutions. Reduction in membrane mineral scaling propensity was evaluated from both flux decline studies (using salt solutions supersaturated with respect to calcium sulfate) and using a transparent RO cell to directly evaluate the time-evolution of surface mineral crystals. The present class of membranes were of higher permeability (by up to a factor of ~2) compared to commercial RO membranes of the same level of salt rejection, but with the added benefit of significantly reduced mineral scaling propensity and comparable fouling resistant when compared with commercial low fouling RO membranes.

See more of this Session: Advances in Water Treatment Membranes by Surface Modification
See more of this Group/Topical: Topical 1: Separation Needs for Energy Independence and Environmental Sustainability

(507b) Modification and Characterization of Regenerated Cellulose Ultrafiltration Membranes for Treatment of Produced Water

Daniel Wandera, Chemical & Biomolecular Engineering, Clemson University, Clemson, SC, Ranil Wickramasinghe, Dept. of Chemical and Biological Engineering, Colorado State University, Fort Collins, CO and Scott M. Husson, Department of Chemical and Biomolecular Engineering, Clemson University, Clemson, SC

Produced water is oily wastewater that is co-produced during oil and gas exploration. Developing a technique that is economically viable and environmentally friendly to treat the large volumes of produced water is a grand challenge for oil and gas companies. Membrane based treatment processes for oily waters often fail due to membrane fouling. We hypothesize that grafting block copolymer nanolayers from commercial membranes will change the surface properties in ways that limit foulant accumulation on these membranes and provide an easy, chemical-free way to remove any attached foulants. Poly(N-isopropylacrylamide)-block-poly(oligoethylene glycol methacrylate) was grafted by surface-initiated atom transfer radical polymerization (ATRP) from two different regenerated cellulose ultrafiltration membranes; a tight 5 kDa membrane and a more open 1000 kDa membrane. The dual functionality provided by this block copolymer system yields fouling resistant and temperature responsive membranes for the treatment of produced water. The temperature responsive block (PNIPAAm) makes it possible for the membranes to self-clean during the filtration of produced water. Fourier-transform infrared spectroscopy spectra confirmed the successful grafting of both polymers from the membrane surfaces. Atomic force microscopy (AFM) and scanning electron microscopy (SEM) were used to characterize changes in the surface topography, and pore morphology of the membranes after modification. We developed synthetic produced water from oil-in-water emulsions for use in this study. The performance of the modified membranes was tested by carrying out water flux measurements using filtration experiments and examining the levels of water flux recovery after a temperature controlled water rinse.

See more of this Session: Advances in Water Treatment Membranes by Surface Modification
See more of this Group/Topical: Topical 1: Separation Needs for Energy Independence and Environmental Sustainability

(507c) Effect of Gradually Varying Baffled-Ring Distance On Ultrafiltration in Tubular Membranes Inserted Concentrically with a Ring Rod

Ho-Ming Yeh, Tamkang University, Tamsui, Taiwan

The decline of permeate flux along the membrane ultrafilters is mainly due to the increment in concentration polarization resistance and the decline of transmembrane pressure. It was found that the concentration polarization resistance could be reduced by the turbulent behavior in a tubular membrane ultrafilter inserted concentrically with a solid rod wrapped with baffle-rings. The performance will be further improved if the baffled-ring distances gradually and properly decrease along the cross-flow channel with fixed number of baffled rings. Theoretical analysis was based on the mass and momentum balances coupled with the application of resistance-in-series model. Correlation predictions are rather acceptable for higher feed concentration comparing with the experimental results for ultrafiltration of dextran T500 aqueous solution in a tubular ceramic membrane of a nine-ring rod insert with gradually decreasing the baffled-ring distances.

See more of this Session: Advances in Water Treatment Membranes by Surface Modification
See more of this Group/Topical: Topical 1: Separation Needs for Energy Independence and Environmental Sustainability

(507d) Ph Responsive Membraens for Water Treatment

Heath Himstedt, Chemical and Biological Engineering, Colorado State Univeristy, Fort Collins, CO, Xianghong Qian, Department of Mechanical Engineering, Colorado State University, Fort Collins, CO and Ranil Wickramasinghe, Dept. of Chemical and Biological Engineering, Colorado State University, Fort Collins, CO

Nanofiltration membranes are frequently used for water treatment. Development of responsive nanofiltration membranes where the membrane properties can be altered by changing environmental conditions could lead to the development of multi-functional membranes. Here nanofiltration membranes have been modified by growing acrylic acid nanobrushes from the surface of the membrane and pores. If the pH of the feed is above the pKa of the grafted nanobrushes, the carboxylic groups will be deprotonated and swell.

Filtration experiments confirm that polyacrylic acid nanobrushes may be grafted from the surface and pores of large pore nanofiltration membranes without significantly impacting the permeate flux. Furthermore, swelling of the grafted nanobrushes at pH values above their pKa leads to a decrease in permeate flux. This results in nanofiltration membranes with properties that can be altered by changing the feed pH. In particular we show how rejection and flux may be varied by adjusting the feed pH. Complimentary modeling studies indicate that the pKa of the tethered polyacrylic acid brushes is affected by graft density and polymer molecular weight.

See more of this Session: Advances in Water Treatment Membranes by Surface Modification
See more of this Group/Topical: Topical 1: Separation Needs for Energy Independence and Environmental Sustainability

(561a) Kinetic Separation of CO_2 and CH_4 On a Copper-Organic Framework

Zongbi Bao, Department of Chemical Engineering, Zhejiang University / New Mexico State University, Las Cruces, NM, Liang Yu, Dept. of Chemical Engineering, New Mexico State University, Las, NM and Shuguang Deng, Chemical Engineering Department, New Mexico State University, Las Cruces, NM

Separation of carbon dioxide from methane is an important issue in the processing of low-quality natural gas. Conventional technology in the natural gas industry uses amines to selectively remove CO_2. However recovery of the CO_2 and solvent is energy-intensive. Pressure swing adsorption (PSA) based on porous materials are widely considered as a more energy-efficient and economical process for CO_2 capture. Considerable attention has been given to the development of PSA processes that are mostly based on differences in adsorption equilibrium.

In this study, we will present the first study attempting to separate CO_2 from CH_4 in a microporous metal-organic framework (MMOF) by kinetic effect. In order to demonstrate the effect of accessible pore size on the kinetic selectivity, we measured diffusion rate of CO_2 and CH_4 in MOF-177 (~10.8 Å), MOF-5 (~7.7 Å), zeolite 5A (~4.4 Å), and Cu(hfipbb)(H_2hfipbb)0.5 (~3.5 Å). Both the adsorption equlibria and kinetics were measured at 278 K, 298 K and 318 K and pressure up to 800 mmHg. Diffusion time constants of CO_2 and CH_4 were investigated from the adsorption uptake curves as a function of time by a micropore diffusion model. An overall selectivity of 20+ was obtained in Cu (hfipbb)(H_2hfipbb)0.5. The result shows this MMOF of interest is a promising adsorbent for adsorptive separation of CO_2/CH_4 mixture.

See more of this Session: Separations Needs for CO2 Capture II
See more of this Group/Topical: Topical 1: Separation Needs for Energy Independence and Environmental Sustainability

(561b) Carbon Dioxide Capture with Activated Carbon: Pilot Unit Test Results

Kaspars Krutkramelis[1], Bryce Dutcher[1], Hertanto Adidharma[2] and Maciej Radosz[1], (1)University of Wyoming, Laramie, WY, (2)Department of Chemical & Petroelum Engineering, University of Wyoming, Laramie, WY

A fully automated carbon dioxide capture pilot unit has been developed for tests on flue-gas slip streams from coal-fired power plants. This process involves a packed-bed sorbent system undergoing two-stage cycles: low-pressure adsorption and steam-aided vacuum regeneration. Early laboratory experiments confirm that 90% recovery and 95% CO2 purity are attainable in a single adsorption/desorption cycle. Field tests are planned this summer to demonstrate that these goals are attainable for over 100 cycles. Following the field tests, economic analysis will be performed to quantify the overall cost per ton of captured CO2 and compare it with alternative CO2 capture technologies.

See more of this Session: Separations Needs for CO2 Capture II
See more of this Group/Topical: Topical 1: Separation Needs for Energy Independence and Environmental Sustainability

(561c) Hollow Fiber-Based Liquid Membrane for CO_2 Capture

Kamalesh K. Sirkar and **Gordana Obuskovic**, Chemical, Biological and Pharmaceutical Engineering, New Jersey Institute of Technology, Newark, NJ

Porous hollow fiber membranes were used to study appropriate immobilized liquid membranes for selective separation of carbon dioxide from anesthesia breathing circuit gas mixture containing nitrous oxide, carbon dioxide, oxygen and halogenated hydrocarbons (HHCs). A number of different membranes and modules were studied at temperatures between 25 – 37 C. Reasonably high selectivities for CO_2/HHC and CO_2/O_2 were observed at room and elevated temperatures (37 C). For CO_2 inlet concentrations of 5%, 2% and 0.5%, the CO_2/O_2 selectivities were in the range of 500, 1200 and 3000 respectively. The CO_2/HHC selectivities were around 30-150+ depending on the CO_2 concentration and the HHC studied.The presentation will describe the utility of these and other novel results for CO_2 capture.

See more of this Session: Separations Needs for CO2 Capture II
See more of this Group/Topical: Topical 1: Separation Needs for Energy Independence and Environmental Sustainability

(561d) Iron-Doped Nickel Catalysts as Anode Materials for An Electrochemical Air Separation Device

James Landon, Chemical Engineering, Carnegie Mellon University, Pittsburgh, PA and John R. Kitchin, Department of Chemical Engineering, Carnegie Mellon University, Pittsburgh, PA

Oxy-fired coal plants as well as integrated gasification combined cycle plants are of interest for their use in capturing carbon dioxide. The production of pure oxygen from the air is essential as a source for these plants. Currently, the technology used to accomplish this air separation process is cryogenic refrigeration. This method is energy intensive and requires a large capital cost. Various other technologies are being proposed to perform this task. One possible route to separate oxygen from the air would be through the use of an electrochemical oxygen concentration cell. These cells can theoretically run with low energy requirements. However, to make this technology viable and approach these low energy costs, catalyst development in the form of more active oxygen reduction and evolution catalysts is necessary. This research investigates more active transition metal oxide catalysts for oxygen evolution. Iron-doped nickel powders were synthesized through an evaporation-induced self-assembly process (EISA) yielding catalysts with BET surface areas ranging from 15-35 m^2/g. Catalysts with doping between 3-10mol% Fe were examined through cyclic voltammetry (CV) and chronoamperometry (CA). It was found that by doping nickel with 5mol% iron, large catalytic gains could be realized. At a modest overpotential of 0.6V, activity of this doped catalyst was increased by 300% over pure nickel oxide. XRD examination of these iron doped catalysts revealed crystalline NiO with no change in the lattice constant suggesting substitutional incorporation of iron into the nickel lattice. Scanning electron microscopy (SEM) and energy dispersive x-ray (EDX) analysis of the electrode surface further showed a well dispersed nickel-iron catalyst. Finally, these catalysts were incorporated into an air separation cell showing successful oxygen separation at potentials lower than 1 V.

See more of this Session: Separations Needs for CO2 Capture II
See more of this Group/Topical: Topical 1: Separation Needs for Energy Independence and Environmental Sustainability

(561e) Entropic Enhancement of CO_2 Capture Using Nanoparticle Ionic Materials (NIMs)

Kunyi Andrew Lin[1], Youngjune Park[1] and Ah-Hyung Alissa Park[2], (1)Earth and Environmental Engineering, Columbia University, New York, NY, (2)Earth and Environmental Engineering & Chemical Engineering, Columbia University, New York, NY

A newly developed material, Nanoparticle Ionic Materials (NIMs), is synthesized with different molecular weight of counter ions (600~2000 polyetheramine) and silica core sizes (7, 12 and 22 nm). A series of analytical tools including TGA, TEM, ATR-IR and NMR are employed to characterize thermal stability, core/counter ions fractions, dispersion, and structural and chemical configuration of synthesized NIMs. It has been hypothesized that the ionic bonds located in corona chains can contribute electrostatic repulsion between corona chains and as a result increase the free volume within the system. Therefore, CO_2 capture by NIMs can be maximized by controlling both enthalpic and entropic effects. In this study, we focus on the CO_2 capture enhancement via the entropic effect through the investigation of the effect of increased free volume on NIMs' CO_2 capture capacity. The experimental setup consists of a temperature and pressure controlled reactor system equipped with a pressure transducer and the effects of temperature (25–65°C) and partial pressure of CO_2 (0.07–0.34MPa) are investigated. It is found that more than 95% of CO_2 absorption in NIMs is achieved during the first 20 minutes and the equilibrium is generally reached in about 50 minutes at 35 °C, P_{CO2} = 0.31 MPa. At the equilibrium, 1 kg of the NIMs sample can capture about 0.16 mole of CO_2. It is also found that temperature and partial pressure of CO_2 can significantly affect the viscosity of NIMs.

See more of this Session: Separations Needs for CO2 Capture II
See more of this Group/Topical: Topical 1: Separation Needs for Energy Independence and Environmental Sustainability

(660a) Hydrophilization of Thin Film Composite Membrane Support Layers for Engineered Osmosis Applications

Jeffrey R. McCutcheon, Department of Chemical, Materials and Biomolecular Engineering, University of Connecticut, Storrs, CT and Jason Arena, Chemical Engineering, University of Connecticut, Storrs, CT

Today's thin film composite membranes have a highly anisotropic structure which provides both high permeability and selectivity in pressure-driven membrane separations like reverse osmosis and nanofiltration. This asymmetric structure, however, is detrimental to engineered osmosis applications, including forward osmosis and pressure retarded osmosis, due to the presence of concentration polarization. The support layers of these membranes are also very hydrophobic. This intrinsic hydrophobicity prevents wetting of the support layers and results in reduced mass transport and enhanced concentration polarization. This investigation considers surface modification with various techniques intended to hydrophilize the support layer of commercial reverse osmosis membranes without causing degradation of membrane performance (selectivity, permeability). These efforts resulted in an increased water flux (by as much as 10-fold) for various EO applications. These efforts may enable the use of commercial reverse osmosis membranes for engineered osmosis.

See more of this Session: Novel Materials for Desalination
See more of this Group/Topical: Topical 1: Separation Needs for Energy Independence and Environmental Sustainability

(660b) Using Nanocomposite Materials Technology to Improve the Properties of RO Membranes

MaryTheresa M. Pendergast and Eric M. V. Hoek, Civil and Environmental Engineering, University of California, Los Angeles, Los Angeles, CA

Composite reverse osmosis (RO) membranes were formed by interfacial polymerization of polyamide thin films over pure polysulfone and nanocomposite-polysulfone porous support membranes. Nanocomposite supports were formed from four different sized amorphous non-porous silica nanoparticles, a crystalline microporous zeolite nanocrystal, and an organic modified form of the zeolite. For each hand-cast membrane, water flux and NaCl rejection were monitored over time for two different applied pressures. Nanocomposite-polysulfone supported RO membranes generally had higher initial permeability and experienced less flux decline due to compaction than pure polysulfone supported membranes. In addition, observed salt rejection generally increased with compaction. Cross-sectional SEM images verified significant reduction in thickness of pure polysulfone supports, whereas nanocomposites better resisted compaction due to enhanced mechanical stability imparted by the nanoparticles. A conceptual model was proposed to explain the mechanistic relationship between support membrane compaction and observed changes in water flux and salt rejection. Nanocomposite-supported RO membranes represent one potential approach to mitigate internal, irreversible fouling due to membrane compaction, particularly in high-pressure applications like seawater desalination.

See more of this Session: Novel Materials for Desalination
See more of this Group/Topical: Topical 1: Separation Needs for Energy Independence and Environmental Sustainability

(660c) Self-Assembled Mesoporous Carbon for Capacitive Deionization of Water

Costas Tsouris[1], David DePaoli[2], Sheng Dai[3], Jim Kiggans[4], Richard Mayes[5], Chia-Hung Hou[6] and Sotira Yiacoumi[6], (1)Oak Ridge National Laboratory, Oak Ridge, TN, (2)Nuclear Science and Technology Divsion, Oak Ridge National Laboratory, Oak Ridge, TN, (3)Chemical Sciences Division, Oak Ridge National Laboratory, Oak Ridge, TN, (4)Materials Science and Technology Division, Oak Ridge National Laboratory, Oak Ridge, TN, (5)Chemical Sciences Division, Oak Ridge National Laboraotry, Oak Ridge, TN, (6)School of Civil and Environmental Engineering, Georgia Institute of Technology, Atlanta, GA

Self-assembled mesoporous carbon materials have been synthesized in our laboratories. Current work aims to translate the discovery of these materials into an industrially viable technology for water desalination via electrosorption or capacitive deionization. Mesoporous carbon sheets have been synthesized and used in a single-cell capacitive deionization reactor. Results have shown that the mesoporous carbon sheets are superior to carbon aerogel materials that are commercially available for capacitive deionization applications. A complex performance has been revealed when a mixture of ions were present in the solution, which is related to the competition of ion size and valence. When sodium chloride was used, ion competition was removed as evidenced from the performance of the reactor. Grand canonical Monte Carlo simulations were performed to study the fundamental aspects of electrosorption selectivity of electrolyte mixtures inside nanopores. Simulation results indicated that exclusion of coions is associated with the population of different counterions involved in the electrical double layer (EDL) formation. In the case of electrolyte mixtures, the competitive effects of asymmetries of ion charge and size can determine the pore accessibility. Divalent counterions have the energetic advantage of preferentially screening the surface charge. On the other hand, small counterions can access the pores and approach closer to the charged surfaces. Electrosorption selectivity of small monovalent versus large divalent ions depends on the magnitude of surface charge. This study is expected to lead to a competitive process for water desalination.

See more of this Session: Novel Materials for Desalination
See more of this Group/Topical: Topical 1: Separation Needs for Energy Independence and Environmental Sustainability

(660d) Fundamental Water and Salt Transport Properties of a Novel Sulfonated Styrenic Pentablock Copolymer

Geoffrey M. Geise[1], Anita J. Hill[2], Benny D. Freeman[1] and Donald R. Paul[1], (1)Deptartment of Chemical Engineering & The Texas Materials Institute, The University of Texas at Austin, Austin, TX, (2)CSIRO Manufacturing Science and Technology, Clayton, VIC, Australia

Sulfonated polymers have been shown to exhibit favorable tolerance to the chlorine compounds that are commonly used to disinfect water. As such, these materials are being studied and optimized for use as chlorine-resistant desalination membrane materials. Fundamental transport studies are needed to understand the structure-property relationships that exist in these materials. These studies will allow further optimization of sulfonated polymers for desalination applications. Here, water and salt transport data have been measured for a series of sulfonated styrenic pentablock copolymers where the degree of sulfonation and the copolymer structure have been varied. The results are presented within the framework of the solution-diffusion model for transport through non-porous materials. In addition, positron annihilation lifetime spectroscopy (PALS) has been used to measure the free-volume characteristics of these materials. Since free-volume is directly related to transport properties in these materials, the PALS technique provides valuable information that can be used to correlate transport properties. Water and salt transport properties will be discussed within the framework of fundamental tradeoff relationships that have been observed in desalination materials. Furthermore, water and salt transport properties will be compared to salt rejection data using traditional reverse osmosis theory.

See more of this Session: Novel Materials for Desalination
See more of this Group/Topical: Topical 1: Separation Needs for Energy Independence and Environmental Sustainability

(696a) Fabrication of Robust Pd-Ag and Pd-Cu Membranes On Microporous Stainless Steel Support by a Novel Electroless Plating Process

Mizanur Rahman[1], Saiful Islam[1], Atikur Rahman[1], Debasish Kuila[2] and **Shamsuddin Ilias**[1], (1) Chemical and Bioengineering, North Carolina A&T State University, Greensboro, NC, (2)Chemistry, North Carolina A&T State University, Greensboro, NC

Dense Pd, Pd-Ag and Pd-Cu composite membranes on miceroporous stainless steel substrate (MPSS) were fabricated by novel electroless plating (EP) process [1]. In conventional Pd-EP process, the oxidation-reduction reactions between Pd-complex and hydrazine result in evolution of NH3 and N2 gas bubbles. When adhered to the substrate surface and in the pores, these gas bubbles hinder uniform Pd-film deposition which results in dendrite growth leading to poor film formation. We addressed this problem by introducing cationic surfactant in the electroless plating process what we call surfactant induced electroless plating (SIEP). The unique features of this innovation provides: (1) control of Pd-deposition rate, and (2) control Pd-grain size distribution. The surfactant molecules play an important role in the plating process in tailoring grain size and the process of agglomeration by removing tiny gas bubbles through adsorption at the gas-liquid interface. The water soluble surfactants which have almost neutral hydrophilic-lipophilic balance (HLB) value are believed to have stronger capability of removing gas bubbles from the substrate surface and the pores. The pre- and post-annealing characterizations of these membranes (Pd, Pd-Ag, and Pd-Cu on MPSS substrate) were carried out by SEM, XRD, EDX, and AFM analysis. SEM images showed significant improvement of the membrane surface morphology, in terms of metal grain structures compared to the membranes fabricated by conventional EP process. From x-sectional SEM and EDX study, we confirmed alloying of Pd-Ag and Pd-Cu at the annealing condition of about 500 C with different annealing time in hydrogen environment. These membranes were also studied for hydrogen perm-selectivity as a function of temperature and feed pressure. Some of these results will be discussed in this presentation.

1. Ilias, S., and Islam, M.A., "Methods for Preparing Thin Films by Electroless Plating." US Patent Application #20100068391, March 18, 2010.

See more of this Session: Membranes for Hydrogen Purification
See more of this Group/Topical: Topical 1: Separation Needs for Energy Independence and Environmental Sustainability

(696b) Pure Silica Sodalite as a Building Block for Hydrogen Separation Membranes

Sanket Shah Champaklal[1], Ying Dai[1] and Vadim Guliants[2], (1)Department of Chemical and Materials Engineering, Univercity of Cincinnati, Cincinnati, OH, (2)Department of Chemical and Materials Engineering, University of Cincinnati, Cincinnati, OH

Hydrogen as a high-quality and clean energy carrier has attracted renewed and ever-increasing attention in recent years, mainly due to developments in fuel cells and environmental pressures including climate-change issues. Hydrogen separation from hydrogen-containing gas mixtures, e.g., syngas, at elevated temperatures is a critical component of the WGS membrane reactor technology. Ultra-microporous sodalite (SOD) is a highly desirable structure for the membrane separation of small gas molecules, such as hydrogen, due to the presence of small (~2.8 Å) six-membered ring openings of the sodalite cages. These openings are accessible only to small molecules such as H2, H2O and He, resulting in a superior membrane performance.

Pure-silica sodalite (Si/Al = ∞) framework is particularly attractive sodalite structure that is hydrophobic in nature, thereby reducing water adsorption and also displaying high chemical and thermal stability due to neutral framework. In this study, we investigated the effect of calcination of pure-silica sodalite crystals as a function of crystal size. IR, SEM/EDS, TGA/DSC, XRD and pore size analysis were performed to study the effects of removal of ethylene glycol occluded during synthesis from the beta cages of pure silica sodalite. The results of this study have important consequences with respect to conditions of seeded synthesis of sodalite membranes that are expected to produce thin, continuous and defect-free sodalite membranes for hydrogen separation at elevated temperatures.

See more of this Session: Membranes for Hydrogen Purification
See more of this Group/Topical: Topical 1: Separation Needs for Energy Independence and Environmental Sustainability

(696c) A Process Dynamic Modeling and Control Framework for Performance Characterization and Enhancement of Pd-Based Membrane Reactors

Reyyan Koc, Nikolaos Kazantzis and Yi Hua Ma, Chemical Engineering, Worcester Polytechnic Institute, Worcester, MA

The syngas produced by coal gasification processes can be utilized in Pd-based water gas shift membrane reactors for the production of pure H_2. Pd/Alloy composite membrane reactors exhibit comparative advantages over traditional packed bed reactors such as simultaneous reaction/separation in one compact unit and increased reaction yields. Within the above context, the present research introduces a comprehensive dynamic process modeling and analysis framework, as well as various process control strategies to further enhance membrane reactor performance resulting in a substantially smaller and functional, safer, environmentally friendlier and more energy efficient process.

In particular, a dynamic modeling framework has been developed mathematically realized by a system of unsteady state material balance equations [1] for both reaction and permeation sides of the membrane reactor in order to analyze and characterize the behavior of a catalytic high temperature water-gas shift Pd-based membrane reactor. In the proposed modeling approach, the original partial differential equations describing the various mass balances were discretized by partitioning the process spatial domain into N sections, for which a well-mixed continuous-stirred tank reactor model is postulated comprised of a system of ordinary differential equations appropriately written as mass balance equations for each component/species [2]. The above modeling approach is particularly suitable and necessary to ensure and facilitate the physical realizability of the proposed process control system. Indeed, proportional integral (PI) controllers were designed and properly tuned in order to induce desirable performance characteristics in the operation of the membrane reactor. Detailed simulation results showed that the membrane reactor reached steady state in 10 s with a CO conversion of 97% exceeding the traditional packed bed conversion levels by ~18%. It should be noted that the reactor's exit stream consisted of 1.9% CO, 72.6% CO_2 and 25.5% H_2 (dry basis) after the implementation of the proper process control strategy.

Two types of feedback control strategies, the servo and regulator, were evaluated [3]. In the servo case, the desired response of the process to a set-point change was enforced by the controller when the CO fraction at the retentate side decreased by 50%. On the other hand, in the regulator case, the total process pressure viewed as a disturbance decreased from 15 to 10 atm unexpectedly. The adverse effect of the decreased total pressure on the overall CO conversion was eliminated with the aid of a PI controller while the CO conversion remained at 97.1 %. Finally, the possibilities of attaining not only extra pure H_2 production objectives, but also CO_2 under high pressure, thus making it amenable to sequestration, were also demonstrated in the context of the present study.

[1] Marcano, J. G. S and Tsotsis, T. T. , "Catalytic Membranes and Membrane Reactors", Weinheim : Wiley-VCH, c2002.

[2] Reyes F. and Luyben W. L., *Ind. Eng. Chem. Res.*, 39 (2000) 3335 – 3346.

[3] Stephanopoulous G., Chemical Process Control, Prentice Hall, Inc, 1984.

See more of this Session: Membranes for Hydrogen Purification
See more of this Group/Topical: Topical 1: Separation Needs for Energy Independence and Environmental Sustainability

(696d) Hydrogen Permeation in Palladium-Based Membranes in the Presence of Carbon Monoxide

Jacopo Catalano[1], Marco Giacinti Baschetti[2] and Giulio C. Sarti[2], (1)Chemical Engineering, Mining and Environmental Technology, University of Bologna, Bologna, Italy, (2)Chemical Engineering, Mining and Environmental technology, University of Bologna, Bologna, Italy

Carbon monoxide is present in steam reforming processes, one of the most promising fields for the application of Pd-based membranes, and it is well known that hydrogen flux through palladium-based membranes is drastically decreased by CO, due to its interaction with the metal surface. In this work a theoretical model to describe hydrogen permeation in palladium and silver-palladium membranes in presence of carbon monoxide is proposed and validated. Unlike the couple of empirical equations proposed to that aim, in the model developed the effects of CO are included in the framework suggested by Ward and Dao whose approach has been modified by considering the competitive adsorption of the different non-inert molecules on the metal interface. The competitive adsorption of CO and H2 has been examined accounting for the spectrum of information available for CO adsorption on palladium, as well as for hydrogen in palladium and palladium-silver alloys. The model has been validated using several literature data spanning in a rather broad range of carbon monoxide partial pressures, temperatures and membrane thicknesses. The data available in the open literature allow the independent estimation of most of the parameters involved in the model so that only one of them needed to be adjusted to obtain a satisfactory representation of the system behaviour, namely the heat of adsorption of CO which is linearly related to the carbon monoxide surface coverage. With the same values of the parameters the model is able to represent well all the different behaviours observed, i.e. the diffusion limited regime obtained with high temperatures and/or large thicknesses, the adsorption limited regime observed with decreasing temperatures in the presence of CO, as well as the desorption limited regime when the effects of CO vanish. The model thus a valuable tool for the prediction of expected membrane behaviour and for membrane modules design. In addition, due to its strong physical bases it can be naturally extended to include the possible presence of other non-inert molecules, also known to affect the H2 permeate fluxes, such as SH2 or water vapor.

See more of this Session: Membranes for Hydrogen Purification
See more of this Group/Topical: Topical 1: Separation Needs for Energy Independence and Environmental Sustainability

(696e) The Effect of Au Content in Composite Pd/Au Membranes Prepared by Galvanic Displacement On Hydrogen Permeance and H2S Resistance

Chao-Huang Chen and Yi Hua Ma, Chemical Engineering, Worcester Polytechnic Institute, Worcester, MA

Pd/Au alloy membranes have received increasing attention recently for separating hydrogen from hydrogen production processes with fossil fuels (e.g. coal gasification) due to the twofold beneficial characteristics. Pd/Au alloy membranes with Au content up to 20wt% showed higher H_2 permeability than pure Pd membranes. In addition, Pd/Au alloy membranes exhibited relatively high sulfur resistance without bulk sulfidation of the Pd/Au layer and less hydrogen permeance decline compared to pure Pd and other Pd alloy membranes in the presence of H_2S. However, the relationship between Au content in composite Pd/Au alloy membranes and hydrogen permeance and sulfur resistance was rarely studied in the literatures. Therefore, the objective of the work was to investigate the effect of the Au content in the composite Pd/Au alloy membranes on hydrogen permeance and the performance in the presence of H_2S. Three composite Pd/Au membranes with different Au content were prepared on 0.1-μ m media grade porous Inconel supports with the electroless plating of Pd until gas tight followed by the Au displacement plating. Prior to the Pd plating, the supports were oxidized (700°C/12h), graded with Al_2O_3, and deposited with Pd/Ag barrier layer for the formation of intermetallic-diffusion barrier layers as well as the modification of the pore size. Three membranes with different Au content were prepared by sequentially depositing different number of Au layers on the membranes (i.e. 1 to 3 Au layers). Between Au layers, a thin Pd layer (0.5 – 0.8 μ m) was deposited for the Au displacement plating. The gravimetric thicknesses of the membranes prepared were 17 - 19 μ m with approximately 4 – 12 wt% Au. After annealing at 500°C in hydrogen, the Pd/Au membranes were characterized in the temperature range of 300 - 450°C in pure hydrogen and helium followed by the sulfur poisoning and recovery tests in the presence of small concentrations H_2S (0.2 - 5ppm) in the temperature range of 400 - 500°C. The Pd/Au membrane of 4 wt% Au showed a hydrogen permeance of 26.8 $m^3/(m^2*h*atm^{0.5})$ at 450°C with an ideal selectivity (J_{H2}/J_{He}, Δp=1 atm) of 4500 after 200 hour testing in pure H_2 and an activation energy of 11.7 kJ/mol. Upon H_2S testing, ~ 62% permeance loss has been observed after a 2 ppm H_2S/H_2 exposure at 450°C. However, the permeance was able to be fully recovered in pure H_2 at 500°C after the poisoning. The hydrogen permeance as well as the performance in H_2S of other Pd/Au membranes with different Au content (i.e. 8wt% and 12wt%) will also be presented with the discussion on the effects of H_2S poisoning on Pd/Au membranes as a function of Au content.

See more of this Session: Membranes for Hydrogen Purification
See more of this Group/Topical: Topical 1: Separation Needs for Energy Independence and Environmental Sustainability

Palladium-Silver Composite Membranes: A Novel Method for Synthesis

Amanda Lewis[1], Sabina Gade[1], J. Douglas Way[1]

[1]*Colorado School of Mines, Golden, CO 80401* alewis@mines.edu, sgade@mines.edu, dway@mines.edu

Inorganic membranes such as palladium (Pd) have promising applications for hydrogen separation and purification [1]. Although Pd is ideal because of its ability to dissociate hydrogen as well as its infinite selectivity for hydrogen, Pd by itself has stability issues [1]. In certain compositions, alloys can have high hydrogen permeability and help alleviate these issues. Pd is alloyed with other metals such as gold (Au), silver (Ag), copper (Cu), etc. One issue is the α to β Pd-H phase transition. This transition expands the lattice which can cause swelling and even cause embrittlement of the membrane. In this work, synthesis of PdAg alloyed membranes is examined due to their ability to lower the hydride phase transition temperature to allow these membranes to operate at lower temperatures [1].

Synthesis of PdAg composite membranes via electroless plating is traditionally done with the bath stabilizer ethylenediaminetetraacetic acid (EDTA) due to the sensitivity of the PdAg co-plating bath to temperature and light [2-4]. EDTA has been shown to remain in the membrane and cause contamination during mixed gas testing [5, 6]. A novel process for the co-plating of PdAg has been developed that does not include EDTA in the electroless plating baths. This process involves the sequential plating of a PdAg layer ($Pd_{80}Ag_{20}$ or $Pd_{70}Ag_{30}$ by weight) followed by a pure Pd layer on top of Praxair Inc. supports. The Praxair Inc. supports are asymmetric, porous zirconia tubes. The PdAg co-plating bath recipes can be seen in Table 1 and the Pd layer plating recipe is the same as Gade, et al [7].

Table 1: PdAg co-plating bath recipes

Solution	Component	Quantity
$Pd_{80}Ag_{20}$	Tetraaminepalladium chloride, 99% pure	1.6 g/L
	Ag nitrate, 99% pure	0.26 g/L
	28-30 wt% Ammonium Hydroxide	432 mL/L
	Deionized Water	568 mL/L
	Hydrazine (1M)	2.5 mL/L
$Pd_{70}Ag_{30}$	Tetraaminepalladium chloride, 99% pure	1.6 g/L
	Ag nitrate, 99% pure	0.45 g/L
	28-30 wt% Ammonium Hydroxide	432 mL/L
	Deionized Water	568 mL/L
	Hydrazine (1M)	2.0 mL/L

The plating sequence is repeated until the film has a room temperature nitrogen leak rate of less than 1.5 mL/min at 20 psi.

Using this novel plating method, three membranes have been made and tested at CSM. Preliminary pure gas permeation testing has shown promising results. Figure 1 is a comparison between the PdAg films and pure Pd films on the same support. The compositions given for each PdAg membrane was estimated from the weight gain of bath compositions.

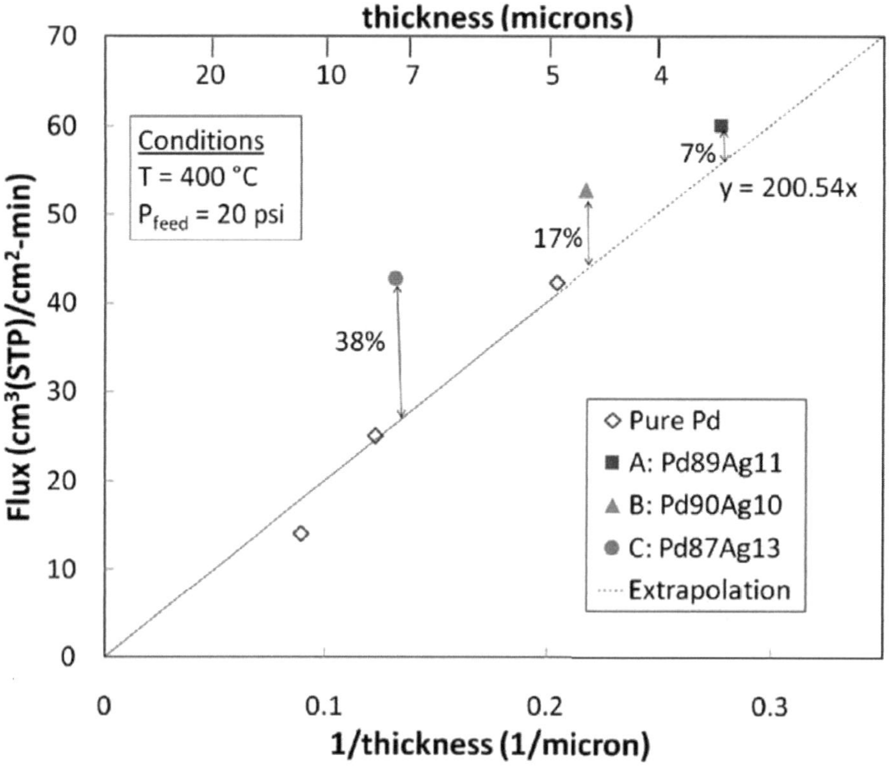

Figure 1: PdAg membranes compared with pure Pd membranes

Given the same testing conditions, all of the PdAg membranes had a higher flux than a pure Pd membrane of a similar thickness. All three of these membranes were annealed differently, but included anywhere from 4 to 20 hours at high temperature (550 °C) under hydrogen. The annealing increased the flux for all three membranes. Membranes A and B have comparable fluxes at 400 °C to a membrane of similar thicknesses and composition as previously reported by Uemiya [8], as seen in Figure 2.

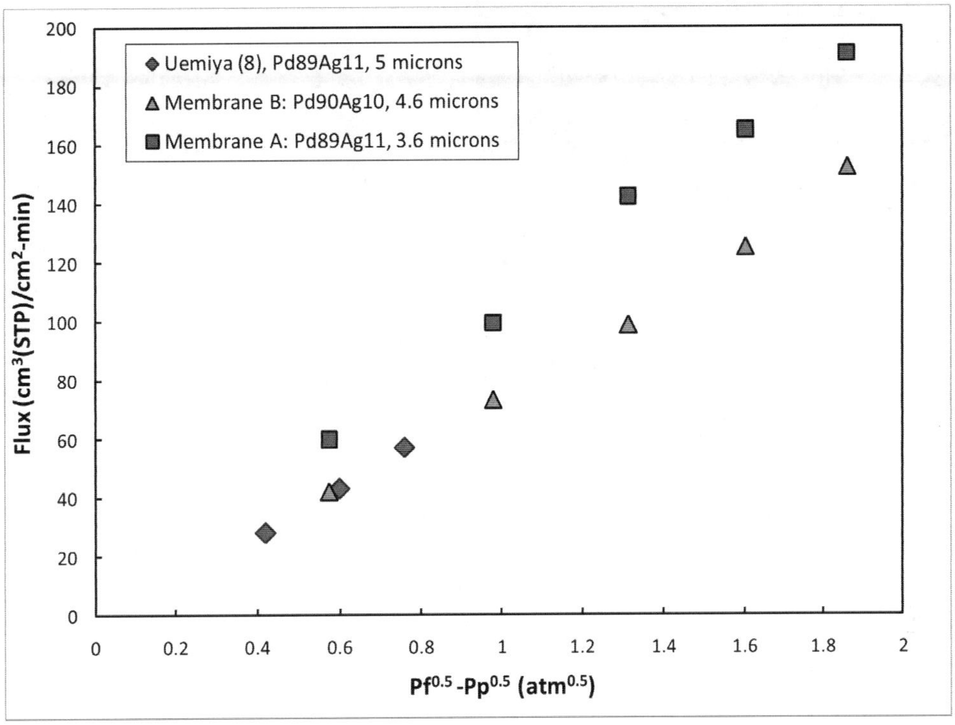

Figure 2: Comparison of CSM membranes to literature at 400 °C

Figure 2 is encouraging because the fabrication technique is able to produce membranes that seem to be annealed based on the comparison of flux values to the literature. After testing, these membranes were cut in half to get an idea if the film was annealed. Figure 3 shows an example of one of the EDS/SEM scans at 2000x. The orange line on the picture on the left side matches up with the orange line on the bottom of the picture of the right side.

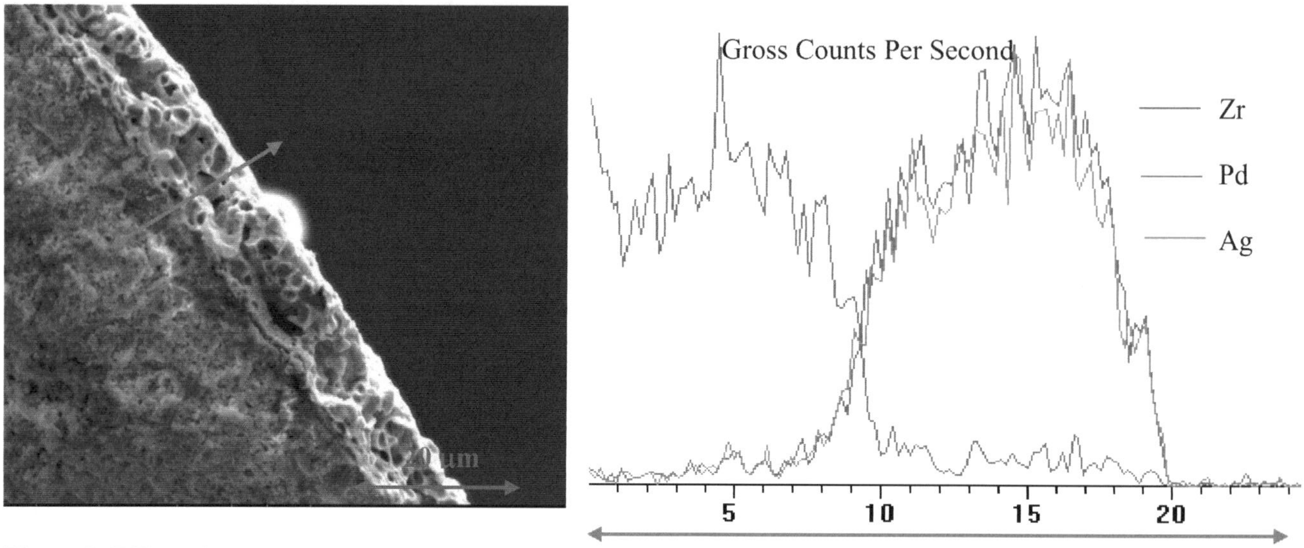

Figure 3: Effect of annealing on PdAg membrane

The scan in Figure 3 shows that the PdAg film is approximately 8-9 microns thick. The zirconia support counts drop dramatically near the surface as the PdAg layers show to be almost on top of each other. This provides good support that the membrane is annealed.

Current work is a twofold focus. The first focal point is to find optimal annealing conditions. During annealing, the thinner membranes nitrogen leak rates begin to creep up. It is unclear if this is due to the membrane seals or pinholes forming in the active area. The second focus area involves the synthesis of the membranes in order to produce thicker membranes and to optimize a plating procedure for better control of the membrane composition. This study is focusing on the $Pd_{80}Ag_{20}$ co-plating bath. The thicker membranes, 5 – 10 μm, will potentially hold up better under annealing as well as during testing. A current goal of the bath optimization is to make a $Pd_{82}Ag_{18}$ membrane in the hopes that it will have very high flux compared to pure Pd and hold selectivity under low temperature (150 °C or less) testing conditions.

This material is based upon work supported by the Department of Energy under Award Number(s) DE-FC26-07NT43054. A huge thank you goes to Praxair Inc. for the ceramic supports and the whole CSM Membrane Research group.

References

1. Paglieri, S.N. and J.D. Way, *Innovations in palladium membrane research.* Separation & Purification Reviews. **31**(1):(2002) p. 1-169.
2. Kikuchi, E. and S. Uemiya, *Preparation of supported thin palladium-silver alloy membranes and their characteristics for hydrogen separation.* Gas Separation & Purification. **5**(4):(1991) p. 261-266.

3. Bhandari, R. and Y.H. Ma, *Pd-Ag membrane synthesis: The electroless and electro-plating conditions and their effect on the deposits morphology.* Journal of Membrane Science. **334**(1-2):(2009) p. 50-63.
4. Witjens, L.C., et al., *Improving the control of the electroless plating synthesis of Pd/Ag membranes for hydrogen separation using Rutherford backscattering.* Journal of Membrane Science. **254**(1-2):(2005) p. 241-248.
5. Collins, J.P., et al., *Catalytic Dehydrogenation of Propane in Hydrogen Permselective Membrane Reactors.* Industrial & Engineering Chemistry Research. **35**(12):(1996) p. 4398-4405.
6. Volpe, M., et al., *Optimized bath for electroless deposition of palladium on amorphous alumina membranes.* Surface and Coatings Technology. **200**(20-21):(2006) p. 5800-5806.
7. Gade, S.K., P.M. Thoen, and J.D. Way, *Unsupported palladium alloy foil membranes fabricated by electroless plating.* Journal of Membrane Science. **316**(1-2):(2008) p. 112-118.
8. Uemiya, S., T. Matsuda, and E. Kikuchi, *Hydrogen permeable palladium-silver alloy membrane supported on porous ceramics.* Journal of Membrane Science. **56**(3):(1991) p. 315-325.

Session: T1007 Membranes for Advanced Fossil Energy Systems

Dense Ceramic Membrane Reactor for CH_4 and CO_2 Utilization

Xueliang Dong, Wanqin Jin* and Nanping Xu

State Key Laboratory of Materials-Oriented Chemical Engineering, Nanjing University of Technology, 5 Xinmofan Road, Nanjing 210009, P. R. China

Methane (CH_4), the main component of natural gas, can be converted to liquid fuels, hydrogen, and other value-added chemicals through a syngas intermediate. Currently, partial oxidation of methane (POM) with pure oxygen in the presence of a catalyst is established to be the most potential process for CH_4 conversion because of its greater selectivity to syngas, its exothermicity, and more desirable CO/H_2 ratio. Meantime, thermal decomposition of carbon dioxide (TDCD) to CO and O_2 ($2CO_2 \rightarrow 2CO + O_2$, $\Delta H^0_{298} = 552 kJ/mol$) is an alternative process for the consumption and utilization of Carbon dioxide (CO_2). However, this reaction is limited by the thermodynamic equilibrium and not easy to be realized in conventional fixed-bed reactors.

In our work, an important advance in using CO_2 as a feedstock for POM process is resulted from the application of a mixed-conducting membrane reactor, which integrated the TDCD and POM processes in a single unit (see Figure 1). [1, 2] TDCD takes place in one side of the membrane and the POM occurs simultaneously in the other side, or more clearly, methane reacts with oxygen, which is permeated through the membrane from the CO_2 decomposition, to produce H_2 and CO over supported transition metal catalysts. At 900 °C, the CO_2 conversion reached about 11.1%.

Then, we developed an innovative catalyst process to decompose CO_2 in an oxygen-permeable membrane reactor packed with a mixed-conducting oxide supported noble metal catalyst, or Pd/$SrCo_{0.4}Fe_{0.5}Zr_{0.1}O_{3-\delta}$ (Pd/SCFZ), which is of

* Corresponding author, E-mail: wqjin@njut.edu.cn. Tel.: +86-25-8317-2266

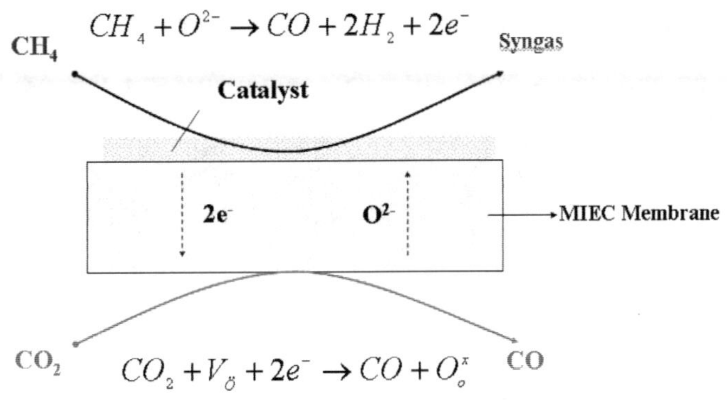

Figure 1. Membrane reactor for coupling thermal decomposition of CO2 with partial oxidation of methane to syngas.

high activity in the decomposition of CO_2 into CO and O_2 [3]. The catalytic activity was expected to be self-regenerated by removing the oxygen from the TDCD reaction by the oxygen permeable membrane. At 900 °C, this catalytic process attains 100% of CO selectivity at 15.8% of CO_2 conversions. Recently, the TDCD reaction was also realized in a thin tubular oxygen-permeable membrane reactor. [4] At 950°C, the CO_2 conversion reached 17.2%, which is higher than the best values reported in literatures so far (see Figure 2).

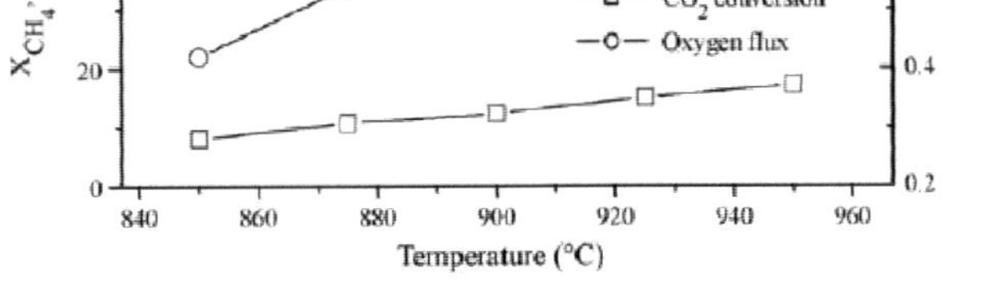

Figure 2. Temperature dependence of CO_2 conversion (X_{CO2}), oxygen flux (J_{O2}), CH4 conversion (X_{CH4}) and CO selectivity (S_{CO}) (feeding conditions-in shell side: 6 cm^3 (STP) min^{-1} CO_2 and 24 cm^3 (STP) min^{-1} He; in tube side: 1 cm^3 (STP) min^{-1} CH_4 and 19 cm^3 (STP) min^{-1} Ar).

The proposed catalytic process can also be extended to the decomposition of other oxygen-containing molecules (such as NO_x and H_2O) and the oxidation of light alkanes (such as ethane and propane). However, we found that the stability of the mixed-conducting membrane reactor was not more than 60h, and the membrane significantly broke due to the erosion of the CO_2 and reducing atmospheres on the surface of the membrane. Moreover, we developed a new kind of $La_{0.85}Ce_{0.1}Ga_{0.3}Fe_{0.65}Al_{0.05}O_{3-\delta}$ (LCGFA) mixed conducting oxide [5], which showed high thermal and chemical stability, low thermal expansion coefficient (TEC), and promising oxygen permeability. Our work indicated that mixed-conducting membrane may play an important role in effective utilization of fossil energy and greenhouse gases.

In our recent work, an electrochemical oxygen pump was used for the decomposition of CO_2. As shown in Figure 3, CO_2 feeds to the cathode side of the oxygen pump and obtains the electron to produce CO and oxygen ion. The oxygen ions transport to anode side through the solid electrolyte and lose electrons to form oxygen molecular.

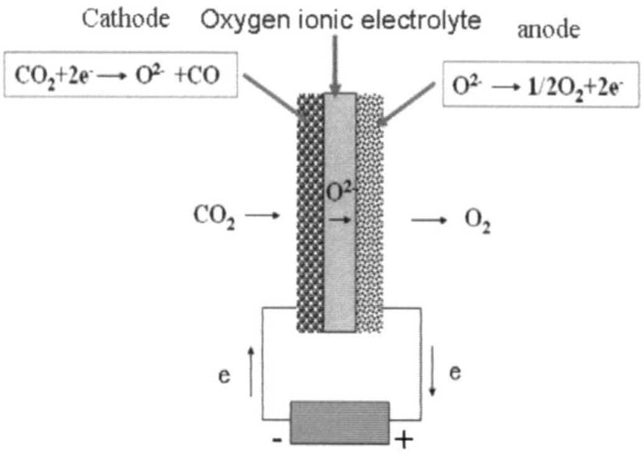

Figure 3. Schematic diagram of the oxygen pump for electrochemical decomposition of CO_2.

$Sm_{0.3}Sr_{0.7}Co_{0.95}Ce0.05O_{3-\delta}$ (SSCC) oxide was used as both cathode and anode materials. $Sm_{0.2}Ce_{0.8}O_{1.9}$ (SDC) was used as the oxygen ionic electrolyte. The cross-section of the electrode was shown in Figure 4. A porous Ag layer, used as a current distributor, can be observed on the surface of the electrode. The second layer is the porous SSCC layer, which was well bonded with both the dense electrolyte layer and the porous Ag layer.

Figure 4. SEM images of the cross section of electrodes.

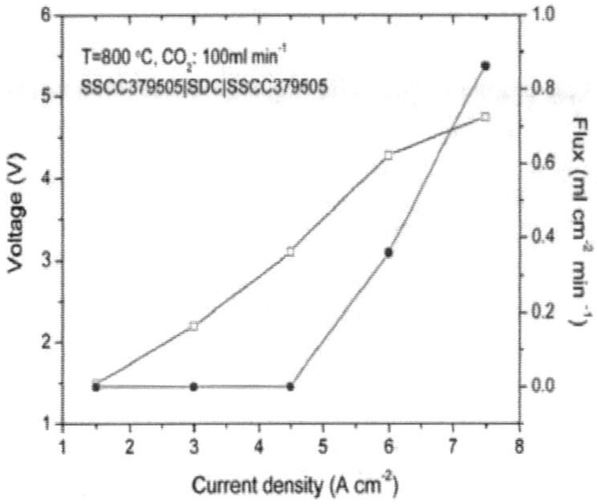

Figure 5 CO_2 electrochemical decomposition performances at 800 °C.

Figure 5 shows the CO_2 electrochemical decomposition performance. No CO_2 decomposition can be observed when the current density was lower than 4.5 A cm^{-2}. The decomposition rate increases with increasing the current density.

Acknowledgements

This work is supported by the National Basic Research Program of China (No. 2009CB623406); National Natural Science Foundation of China (No. 20990222) and China Postdoctoral Science Foundation funded project (No. 20090461105).

Reference

[1] Jin, W. Q.; Zhang, C.; Zhang, P.; Fan, Y. Q.; Xu, N. P. *AIChE J.* **2006,** 52, 2545–2550.

[2] Zhang, C.; Chang, X. F.; Fan, Y. Q.; Jin, W. Q.; Xu, N. P. *Ind. Eng. Chem. Res.* **2007,** 46, 2000–2005.

[3] Jin, W. Q.; Zhang, C.; Chang, X. F.; Fan, Y. Q.; Xing, W. H.; Xu, N. P. *Environ. Sci. Technol.* **2008,** 42, 3064–3068.

[4] Zhang, C.; Jin, W. Q.; Yang, C.; Xu, N. P. *Catal.Today* **2009**, 148, 298–302.

[5] Dong, X. L.; Zhang, G.R.; Liu, Z. K.; Zhong, Z. X.; Jin W. Q.; Xu, N. P. *J. Membr. Sci.* 2009, 340, 141–147.

(732b) Effect of Exposure to Water Gas Shift Gases Containing Sulfur On the Performance of Palladium Alloy Membranes

Stephen N. Paglieri[1], Sarah J. DeVoss[1], Gökhan O. Alptekin[1], Oyvind Hatlevik[2], Sabina K. Gade[3] and J. Douglas Way[3], (1)TDA Research, Inc., Wheat Ridge, CO, (2)Department of Chemical Engineering, Colorado School of Mines, Golden, CO, (3)Colorado School of Mines, Golden, CO

Hydrogen derived from fossil fuel and biomass resources has to be separated and purified for it to be used in fuel cells and turbines. Metal membranes must be able to efficiently purify hydrogen at the relatively high temperatures and pressures typical of catalytically steam reformed gasified coal and biomass. Membranes must also be able to withstand long-term operation in the presence of common constituents of water-gas shift (WGS) gases including CO and H_2S. Using electroless plating techniques, pinhole-free Pd–Au membranes less than 10 μ m thick were deposited onto AccuSep® (Pall Corp.) porous stainless steel supports coated with a yttria-stabilized zirconia metallic interdiffusion barrier. Pd–Au membranes containing up to 23 wt.% Au were prepared with a range of thicknesses (1.7-9.8 microns). The supported thin film Pd–Au membranes were tested at 400°C in different concentrations of simulated WGS gases containing H_2, H_2O, CO, CO_2, and 20 ppm H_2S in tests lasting for up to 120 hours. For example, a 3.6-micron thick Pd-5.5wt.%Au membrane had a hydrogen flux of 1.36 mol m^2 s^{-1} in a mixture of 50% H_2, 30% CO_2, 19% H_2O and 1% CO at a transmembrane pressure differential of 11.4 bar. The hydrogen purity was 99.993%. Pd–Au alloys displayed stable hydrogen flux during exposure to WGS gases without sulfur. Greater resistance to sulfur was observed with increasing Au content and the membranes recovered a majority of their original H_2 flux after testing in WGS mixtures containing 20 ppm H_2S. However, exposure to H_2S eventually resulted in a decrease in hydrogen selectivity of the composite membranes. The performance of the Pd–Au composite membranes will be compared to results using self-supported Pd–Au alloy foils.

See more of this Session: Membranes for Advanced Fossil Energy Systems
See more of this Group/Topical: Topical 1: Separation Needs for Energy Independence and Environmental Sustainability

(732c) Modeling and Analysis of Zeolite Membrane Reactor for Water Gas Shift Reaction for Hydrogen Production

Zebao Rui[1], Haibing Wang[1], Jerry Y.S. Lin[1] and Junhang Dong[2], (1)School of Mechanical, Aerospace, Chemical and Materials Engineering, Arizona State University, Tempe, AZ, (2)Chemical Engineering, University of Cincinnati, Cincinnati, OH

Water gas shift reaction (WGS) is an important step in hydrogen production from fossil fuels. Due to its exothermic nature, it is known that the reaction tends to shift to the left side at high temperatures, which limits the conversion of the reaction. Membrane reactors, which integrate reaction and separation into the same unit, have the ability to improve the conversion and overcome the thermodynamic limit due to the ongoing selective removal of the product during reaction. For the commercial application of the high temperature (>400 oC) WGS process, the membrane reactor needs to have sulfur tolerance and hydrothermal stability. Crystalline microporous zeolite membranes, especially the siliceous MFI-type zeolite membranes, can meet these requirements [1]. The primary mass transport channels in the MFI-type zeolites have an effective diameter of 0.56 nm, which offer high selectivity by size discrimination for critically sized molecules. The latest results reported by Tang et al. [2] showed that the H2 selectivity was dramatically enhanced at the cost of a moderate decrease in the H2 permeance of the ƒÑ-alumina supported siliceous MFI-type zeolite membranes by selective depositing molecular silica in the zeolitic channels via in-situ catalytic cracking deposition (CCD) of silane precursor. Additionally, the modified membrane exhibited good stability in the presence of water vapor at temperatures >400 oC because of the chemical bonds between the silicious molecular deposits and the zeolite surface [2]. These promising experimental results show that the modified MFI-type membranes offer a great potential to construct a membrane reactor for WGS. In this work, the potential and the features of the WGS MFI-type membrane reactor will be analyzed by a modeling method.

An isothermal steady-state model has been developed for simulating WGS in an isothermal tube-shell MFI-type zeolite membrane reactor. The model considers the permeation properties of the MFI membrane measured in the literature [2] as a reference, and the kinetic equation developed for the Fe-Cr-O based catalyst [3] for WGS reaction. The effect of the membrane reactor design and operational parameters, such as pressure, temperature, inlet space rate, steam to CO ratio (S/C), and the relative values of membrane permselectivity and permeation flux, on the performance of a WGS membrane reactor has been studied. The performance was evaluated based on the CO conversion and H2 recovery.

The results show that the improvements in CO conversion and H2 recovery can be achieved with the reference permeation properties. For a space velocity of 10,000 h-1, with a steam to CO ratio of 1 and pressure difference between the reaction side and the sweep side of 30/1 atm, a CO conversion of 92% and H2 recovery of 97% can be obtained at 500oC. A pressure increase allows for the attainment of a higher conversion at the same space time. The increase of S/C has a negative effect on both the reaction rate and the gas fluxes due to its dilution effect when operating under kinetic limited conversion conditions with a constant CO inlet flow rate. When operating under equilibrium limited conversion conditions, high S/C values allow a better performance to be obtained in terms of reaction but worse preformance in terms of separation when performed at a low pressure (1.5/1 atm). When operating at a high pressure (30/1 atm), there is no obvious advantages of a high S/C ratio even under equilibrium limited conditions.

Keywords: Modeling and simulation; WGS; Zeolite membrane reactor; Hydrogen production

Reference [1] J. Dong, Y.S. Lin, M. Kanezashi, Z. Tang, J. Appl. Phys. 104 (2008) 121301. [2] Z. Tang, T.M. Nenoff, J. Dong, Langmuir, 25 (2009) 4848. [3] W.F. Podolski, Y.G. Kins, Ind. Eng. Chem. Process Des. Dev. 13 (1974) 414.

See more of this Session: Membranes for Advanced Fossil Energy Systems
See more of this Group/Topical: Topical 1: Separation Needs for Energy Independence and Environmental Sustainability

(732d) Polymer/Imidazolate Framework and Polymer/Silica Sodalite Mixed Matrix Membranes for Gas Separations

Ying Dai, Oxana V. Fursenko, Sanket Shah Champaklal and Vadim V. Guliants, Department of Chemical and Materials Engineering, University of Cincinnati, Cincinnati, OH

Mixed matrix membranes (MMMs) for gas separations are traditionally comprised of zeolites dispersed in a polymer matrix in order to improve desired penetrant permeability and selectivity from polymer membranes. Zeolites have a variety of drawbacks for this application – chiefly, a limited number of possible zeolite structures and hydrophilic inorganic surface – these limit their potential for use in MMMs. Zeolitic imidazolate frameworks (ZIFs) are new class of microporous materials. They are an alternative to the use of zeolites in MMMs because the number of possible structures is effectively infinite. This work describes the preparation of polymer-based mixed matrix membranes (MMMs) by casting from polysulfone (PSf) and polyimide (Ultem) chloroform solutions containing dispersed ZIFs or silica sodalite. The membrane microstructure and transport properties were investigated by SEM, X-ray diffraction, gas adsorption and permeation measurements. The gas transport properties of these MMMs with ZIFs and silica sodalite showed improvement over the pure polymer gas transport properties in both selectivity and permeability. For example, when ZIF-8/polyimide (Ultem) mixed-matrix membranes were studied, the He/CO2 selectivity increased from 7 to 12 and the permeability of He increased from 9.4 to 12.6 Barrers. The increased He/CO2 selectivity in ZIF-8/polyimide (Ultem) membranes was explained by the sieving effect introduced by the pure ZIF-8 crystals (pore window 0.34 nm) that restricted the transport of molecules larger than He.

See more of this Session: Membranes for Advanced Fossil Energy Systems
See more of this Group/Topical: Topical 1: Separation Needs for Energy Independence and Environmental Sustainability

(732e) A Novel Au/NaY Catalytic Membrane for Preferential Oxidation of CO From Hydrogen

Xuehong Gu[1], Ziping Zhu[2] and Nanping Xu[2], (1)College of Chemistry and Chemical Engineering, State Key Laboratory of Materials-Oriented Chemical Engineering, Nanjing University of Technology, Nanjing, China, (2)State Key Laboratory of Materials-Oriented Chemical Engineering, Nanjing University of Technology, Nanjing, China

Proton exchange membrane fuel cells (PEMFC) have several advantageous such as its low operating temperature and high energy efficiency. However, the Pt anodes are easily poisoned by CO. It is critical to eliminate CO from the H2-rich gas stream before fed into PEMFC. Preferential oxidation of CO (CO-PROX) is considered as a simple and economic method for CO removal. Over the past few years, the observation in striking catalytic activation of nanosized gold towards CO selective oxidation has motivated many researchers working on the new catalysts. Many efforts were made to improve the catalytic efficiency for CO-PROX. In this paper, we developed a novel catalytic membrane for CO-PROX, in which a polycrystalline NaY (FAU-type) zeolite membrane supported on a porous ceramic substrate was loaded with nanosized Au particles. When H2-rich gas containing a small amount of CO diffuses through the catalytic membrane, CO could be preferentially oxidized due to the selective catalytic activation for Au particles and relatively low diffusion rate for CO molecules. The catalytic configuration was expected to have high efficiency because the reactive components can well reach the activation sites. We have demonstrated that the catalytic membranes had excellent catalytic performance for selective oxidation of CO. CO/H2 molar ratio in the permeate side decreased with the increasing of operating temperature in the range of 80-200 ^{o}C. At 200 ^{o}C, almost no CO was detected from the permeate stream of a catalytic membrane for the feed containing 0.67% CO, 1.33% O2, 32.67% H2 and He in balance.

See more of this Session: Membranes for Advanced Fossil Energy Systems
See more of this Group/Topical: Topical 1: Separation Needs for Energy Independence and Environmental Sustainability

(732f) Development of Nanocrystalline Thin Films of Proton Conducting Perovskite Oxides for High Temperature H2 Separation and Optical Probing in Fossil Fuel Gases

Xiling Tang[1], Zhi Xu[1], Junhang Dong[1], Hongbiao Duan[2] and Hai Xiao[2], (1)Department of Chemical and Materials Engineering, University of Cincinnati, Cincinnati, OH, (2)Department of Electrical and Computer Engineering, Missouri University of Science and Technology, Rolla, MO

This study focuses on the synthesis and characterization of nanocrystalline thin films of proton conducting perovskite oxide Sr(Ce0.8Zr0.1)Y0.1O2.95 (SCZY) thin films and their chemical and physical properties. The SCZY thin films have been coated on various substrates including Si wafers, sapphire chips, and micro-sized optical fibers. The H2 sorption uptake, electrical conductivity, and optical response to H2 containing gases have been experimentally measured as a function of temperature. The potential utilization of the SCZY nanofilms as high temperature H2 separation membranes and optical probes for H2 in fuel gases are dicussed.

See more of this Session: Membranes for Advanced Fossil Energy Systems
See more of this Group/Topical: Topical 1: Separation Needs for Energy Independence and Environmental Sustainability

AUTHOR INDEX

AUTHOR INDEX

American Institute of Chemical Engineers
3 Park Avenue
New York, New York 10016-5991

CURRAN ASSOCIATES INC
proceedings.com